Will Robots Take Your Job?

New Human Frontiers series

Nigel M. de S. Cameron, *Will Robots Take Your Job?*
Harry Collins, *Are We All Scientific Experts Now?*
Everett Carl Dolman, *Can Science End War?*
Mike Hulme, *Can Science Fix Climate Change?*
Margaret Lock & Gisli Palsson,
Can Science Resolve the Nature/Nurture Debate?
Hugh Pennington, *Have Bacteria Won?*
Hilary Rose & Steven Rose, *Can Neuroscience
Change Our Minds?*

Will Robots Take Your Job?

A Plea for Consensus

NIGEL M. DE S. CAMERON

polity

First published in 2017 by Polity Press

Polity Press
65 Bridge Street
Cambridge CB2 1UR, UK

Polity Press
350 Main Street
Malden, MA 02148, USA

ISBN-13: 978-1-5095-0955-3
ISBN-13: 978-1-5095-0956-0(pb)

A catalogue record for this book is available from the British Library.

Library of Congress Cataloging-in-Publication Data

Names: Cameron, Nigel M. de S., author.
Title: Will robots take your job? : a plea for consensus / Nigel M. de S. Cameron.
Description: Malden, MA : Polity, 2017. | Series: New human frontiers | Includes
    bibliographical references and index.
Identifiers: LCCN 2016049516 (print) | LCCN 2016057625 (ebook) | ISBN
    9781509509553 (hardback) | ISBN 9781509509560 (paperback) | ISBN
    9781509509584 (Mobi) | ISBN 9781509509591 (Epub)
Subjects: LCSH: Labor supply–Effect of technological innovations on. |
    Technological innovations–Social aspects. | Automation. | BISAC: SCIENCE /
    Philosophy & Social Aspects.
Classification: LCC HD6331 .C267 2017 (print) | LCC HD6331 (ebook) | DOC
    331.12–dc23
LC record available at https://lccn.loc.gov/2016049516

Typeset in 11 on 15 pt Adobe Garamond Pro Regular
by Toppan Best-set Premedia Limited
Printed and bound in Great Britain by Clays Ltd. St. Ives PLC

For further information on Polity, visit our website: politybooks.com

For three little boys, my youngest grandchildren, who I trust will lead long and blessed lives among the machines in the world of their economic possibilities: Euan Cameron, Gideon Robison, and Lincoln Balmer.

We are being afflicted with a new disease of which some readers may not yet have heard the name, but of which they will hear a great deal in the years to come – namely, technological unemployment. This means unemployment due to our discovery of means of economising the use of labour outrunning the pace at which we can find new uses for labour.

John Maynard Keynes (1931)

Exactly how bad is the job situation going to be? An Organization for Economic Cooperation and Development study concluded that 9% of American jobs are at risk. Two Oxford scholars estimate that as many as 47% of American jobs are at risk. Even the optimistic scenario portends a serious problem. Whatever the case, it will need to be possible, within a few decades, for a life well lived in the U.S. not to involve a job as traditionally defined... The case for "this time is different" has a lot going for it.

Charles Murray (2016)

CONTENTS

Preface   ix
Acknowledgments   xv

Introduction: Time to Stop Being Naive   1

1 Non-Human Resources   14

2 "The Stupid Luddite People"   43

3 Welcome to the Rust Belt   60

4 Building Consensus and Getting Prepared   87

Notes   110
Bibliography   112
Index   117

A short book on a complex and controversial topic is a rather dangerous thing to write. But the questions addressed here, at the meeting point of science, technology, and society, are questions for all of us, and we can't avoid them.

I've been concerned about the question of the future of jobs for many years, and in the past have made efforts to convene discussions among policymakers and other leaders. Until three or four years back hardly anyone was prepared to talk; and despite the fact that there is now a discussion taking place, at least on the fringe of public life (evident in the appearance of a spate of new books, and quotable quotes from the likes of Bill Gates, Larry Summers and Charles Murray), things have not changed that much. Political leaders still skirt the question and its enormous implications.

My personal interactions have been instructive.

A few years ago, I sat down for lunch on a sunny Menlo Park day with two partners from a global law firm who work with clients in Silicon Valley. There were various things on my mind, from the wine to how we get Washington more focused on innovation. But that was not the conversation they had planned. "When," asked the top guy, "are we going to come up with innovations that create jobs rather than destroy them?" You could have knocked me down with a chip.

Shortly after, I was back in Washington and having dinner beside a top official in the US Department of Labor. I asked who on her team was working on this issue. No one, she said, and changed the subject.

The Washington think tank I lead convened a discussion some time back that involved AFL-CIO (the US labor union network) and a leading government economist, together with technology guru Marshall Brain. I asked someone from AFL-CIO who was working on this issue. When he replied no one, I gently suggested that perhaps they should all stop what they were doing and re-focus here. When we did convene the discussion, the labor union participant seemed little concerned. I tried to goad him by talking about self-driving trucks. What if, in ten or fifteen years' time, the Teamsters, the truck drivers' union, end up with precisely no members? Soon after, I was involved when a UK institute took a similar initiative in London. Once again the labor union point of view was very much in line with the conventional wisdom. It seems that even labor leaders are worried about being labeled Luddites.

More recently, an invitation came to address a group of economists at the World Bank. I walked into their den with trepidation. What I found interesting was that after I'd finished no one informed me that "economics" tells us all will be well (as we sometimes hear from the techno-boosters). Of course, as we shall see, some of the most important economists in history have been among those raising the alarm. Slowly, the issue is creeping into mainstream discussion.

Again recently, I was invited to join in a fascinating conference on what the world will be like in 100 years' time. The letter of invitation, from former Brazilian President Cardoso, made the beguiling point that a child born today may well be alive in 100 years' time. Those of us with children or grandchildren need to think a very long way ahead if we are to focus on their interests and likely futures. There is nothing "academic" about the question. What do I advise Euan, and Lincoln, and Gideon, and my other grandchildren to plan for in their future careers? We discuss these issues as academics, thought leaders, politicians, business people, technologists, civil society representatives. But also as workers ourselves. And as parents and grandparents.

What does the future hold for our children, and their children?

We have been raised, those of us in the developed world, at a time of unprecedented prosperity. Our communities include poor people as well as those who are better off. And our global community includes hundreds of millions who remain extremely poor as well as broad disparities between those who merely get by and those who have recently been called the 1 percent. What are the implications of the coming disruption for the grave economic divides that currently slice up the human race into groups of haves and have-nots?

We look to the prospect of a prosperous future for the global community, powered by the amazing combination of hard work and imagination that has brought us from primitive times through agricultural and industrial societies to the brink of a new world in which digital technologies are set to drive a fresh economic and social order for the species. No question, the core element in that new prosperity will be the relief of many from the laborious work that has in past centuries been our only human way to sustain ourselves. First we made basic but powerful tools, then we harnessed animals, then the Industrial Revolution engaged steam power and a rash of inventive machines – and the rest is indeed history: industrial development spreading around the globe, and the very rapid extension of the digital revolution, enabling near-universal access to knowledge through the internet and bringing hundreds of millions of still-poor people into the global knowledge economy through mobile technology.

And so, the development of a new tech "species," our latest tool, offers to lift the burden of work from our shoulders and deliver unprecedented prosperity. Perhaps the pace and complexity with which it develops will leave us, increasingly, with little to do, as we move into what Norbert Wiener memorably termed a "slave economy." With the fading of the significance of labor in producing value, the dramatic shift of economic power to the owners of capital will test the capacity of our democracies to develop new models of distribution. It may also lead us into a wholly unprecedented situation, depicted by Keynes as the emergence of a "new leisured class," with fundamental challenges for human society as our time is freed up for activities no longer tied to earning our daily bread.

Of course, this may be an exaggeration, or describe a state of affairs that lies only far in the future. But what seems clear, even on a cautious reading, is that labor markets will face turbulence as waves of "rust belt" structural unemployment hit jobs built on skill sets at both the top and the bottom of the scale, and as, we trust, new jobs emerge, forcing workers to adapt and retrain. What also seems clear is that there is a risk, a non-trivial risk – one that thinkers from Larry Summers to Charles Murray have begun to flag, and anticipated by economists from Ricardo to Keynes – that while new jobs will undoubtedly emerge we are in the process of a fundamental shift that will leave the "full-employment" economy behind. We need to reflect on the implications of such a potential shift for labor markets, income and wealth, and for our use of time. Whatever our personal take on the likelihood of such a prospect, we need to make our best efforts to be prepared.

While that's true of us as individuals, a special weight of responsibility rests on the shoulders of the leaders, political and other, who take decisions on our behalf and that of the global community. Their absence from this discussion is remarkable and somewhat alarming. It needs to end.

Nigel M. de S. Cameron
Washington, DC
September 24, 2016

ACKNOWLEDGMENTS

It's always invidious to single out colleagues for thanks when one has had so many fruitful conversations on the subject, but let me single out my C-PET friends and colleagues Nagy Hanna, Jenn Sertl, and Matt James, with whom I have had many. I've also mentioned in the text various individuals who have helped shape my thinking along the way.

Needless to say, like other colleagues in the Center for Policy on Emerging Technologies network, they bear no responsibility for the views I have expressed – although some have helped sharpen the arguments and all have contributed to my thinking since C-PET was established back in 2007. Ana Goelzer invited me to speak on this matter at a TEDx in Porto Alegre, Brazil. Most recently, I'm grateful to my colleagues at the University of Ottawa's Institute for Science, Society and Policy – director Monica Gattinger in particular – where I was able to work on this project for part of my time as Fulbright Visiting Research Chair in Science and Society in 2015–16.

look up

Acknowledgments

Finally, my special thanks to Jonathan Skerrett of Polity Press for approaching me with the suggestion for this book and for his fastidious care in its production. You will decide whether it was a good idea.

Time to Stop Being Naive

Economists have long reflected on the impact of technology on employment, sometimes with anxiety, but often with the confidence that there will always be plenty of jobs. Recent, rapid advances in robotics and AI are forcing the issue. How can we make sense of something on which there is so much disagreement?

> Software substitution, whether it's for drivers or waiters or nurses . . . it's progressing. Technology over time will reduce demand for jobs, particularly at the lower end of the skill set . . . 20 years from now, labor demand for lots of skill sets will be substantially lower. I don't think people have that in their mental model.
>
> Bill Gates, speech to the American Enterprise Institute, 2014

As I write, a remarkable experiment has just begun in the American city of Pittsburgh. Uber, the app-based ride-hailing service that has challenged taxi cabs and given work to hundreds of thousands of amateur drivers, is now offering a self-driving car service (Dwoskin & Fung, 2016). What was unthinkable only a decade ago, and has been experimental since, is now in commercial operation. What does this signify for the relations between humans, machines and jobs?

We're at the outset of a great debate. At one level, it's simple. It's about whether we need to worry that robots will take our jobs, or whether we don't. It's plainly an enormously important debate, with implications for humankind that could be as great as the implications of climate change. But, so far, it has taken place on the fringes of public conversation, and essentially been ignored by our leaders.

The conventional wisdom has been that we don't need to worry, and some technology proponents have argued for this vociferously. At the same time, there are famous thinkers – including John Maynard Keynes, the most influential economist of the last 100 years, and Norbert Wiener, the acclaimed "father of cybernetics" – who have suggested that the rise of Machine Intelligence leading to the collapse of human employment is a serious possibility. It's true that our experience so far of new technologies has been such that, while they may cause disruption to individuals and to industries, they also lead to new jobs – and greater wealth. But will this be true of our latest invention – machines powered by Artificial Intelligence, designed specifically to perform tasks of many kinds that up till now have given us humans employment?

Of course, there aren't simply two views in the debate, but a whole range. Some people predict that in coming decades most human work will just disappear. At the other extreme, there are those who accuse anyone who even raises the issue of standing in the way of progress and acting like a Luddite.[1] In between lies the reasonable expectation that, even if new jobs do emerge in large numbers to replace those that go to machines, the disruption involved will mimic that of earlier industrial shifts – from the Industrial Revolution itself to the much more recent collapse of heavy manufacturing in the US and UK that led to the "rust belt," and a long and painful transition for the workers involved. But, when all's said and done, given that technological change only affects us within a social context (shaped by responses from government and the legal system), we face the question: how are we preparing for these eventualities?

While none of us knows what's going to happen, we do know that if there is a big disruption coming in the labor market we should be worried. Modern societies run on the assumption of what economists call "full employment," which does not mean that everyone has a full-time job but that most people who want one can get one. Will this still be true, in ten, twenty, or thirty years' time? The stakes could hardly be higher.

Since we don't *know* what's going to happen, we can summarize the situation we face in two questions: 1) *On the conventional assumption that new jobs will emerge to take the place of those that go to machines, what kind of labor market turbulence can we expect during the transitions that will be involved?* 2) *Is the idea of big job losses that aren't compensated for with new jobs just ridiculous, or a serious possibility?* Despite some academic debate and occasional involvement from tech and other opinion leaders, there's no evidence to suggest that these questions are being taken seriously by governments. Plainly, if the answer to Question 1 is "not much labor market turbulence," and to Question 2, "yes, the notion of big net job losses is just ridiculous," then we've no need to worry. The argument of this book is that while these are both very comforting responses, they are unreasonable. And they are potentially dangerous, because they wrongly assess the risks we face going forward.

But how do we decide our responses? When we personally "don't know," and when the issue is in fact rather complicated, one approach may be to look at whether there are smart people out there who have expressed their own opinions – and in the process, as it were, opened the door for us to join the conversation.

Here are some examples. I've already mentioned the great John Maynard Keynes, writing as far back as the 1930s, and, working around the same time, Norbert Wiener. But let's get up to date. Bill Gates, in the epigraph to this chapter, has put it like this: Demand for labor is going to go down, and "We don't have it in our mental model." Strikingly, two of America's most prominent public intellectuals have also recently stepped into the arena. Larry Summers, former Secretary of the Treasury (and President of Harvard), from the liberal end of the political spectrum, has undergone a conversion experience and now believes the "Luddites" may be right (Summers, 2013), while the conservative intellectual Charles Murray, writing in the *Wall Street Journal* and under the aegis of the pro-business American Enterprise Institute, sees the upcoming turbulence in labor markets and new employment patterns as part of his case for a Universal Basic Income (Murray, 2016).

The Pew Research Center recently ran an interesting project focused not on researching jobs themselves, but on what "experts" think is likely to happen (Smith & Anderson, 2014). It's no surprise to find the experts divided, though the extent of their division is striking.

In answer to the question, "Will networked, automated, artificial intelligence applications and robotic devices have displaced more jobs than they have created by 2025?," the respondents, of which there were over 2,000, split 48 percent in favor of this assessment to 52 percent against.

Commenting on this result, *Harvard Business Review* writer Walter Frick (2014) wrote: "To those in fear of being replaced by automation, the fact that experts are divided may seem like consolation – unfortunately, it's anything but." Frick goes on to cite another poll (conducted by the University of Chicago's business school) which asked a sample of economists for their take on the impact of *past* automation. Only 2 percent actually disagreed with the statement that "Advancing automation has not historically reduced employment in the United States." In other words, looking back, economists are near unanimous in going with the "conventional wisdom" that "while technology may displace workers in the short-term, it does not reduce employment over the long-term." The fact that the Pew poll's result is dramatically different when experts are asked to look forward "signals the recognition that this wave of technological disruption could in fact be different." Which brings us to the marrow of the issue: *Is it different this time?* And if so, why might that be the case? For one thing, shifts in training and skills development have in the past had a lot more time to work through the labor force. Futurist Bryan Alexander, one of the Pew respondents in the report, writes as follows: "The education system is not well positioned to transform itself to shape graduates who can 'race against the machines.' Not in time, and not at scale. Autodidacts will do well, as they always have done, but the broad masses of people are being prepared for the wrong economy" (quoted in Smith & Anderson 2014).

A related point in the Pew Report comes from science writer and *Economist* magazine deputy editor Tom Standage: "Previous technological revolutions happened much more slowly, so people had longer to retrain, and [also] moved people from one kind of unskilled work to another. Robots and AI threaten to make even some kinds of skilled work obsolete." So why is there not a much more serious and high-profile conversation taking place on these questions?

Partly because, if you raise the issue, someone is likely to call you names – specifically, to invoke the mythical "Ned Ludd." Nobody wants to be called a Luddite.

Partly because it's complicated. Technologies are difficult to understand, and their impact is unpredictable. What's more, there's no obvious location on the political spectrum for the discussion either – and no aspiring leader, from the left or the right, wants to be seen as anti-technology.

Partly because it's a long-term discussion, and politics tends to be short term.

Yet smart people know that disruption is never all good news, and the impacts of the digital revolution have us riding a huge disruptive wave. At the apex of that revolution lie rapid advances in Artificial Intelligence. We've barely begun to explore its potential. But from a policy perspective, and a personal one too, head-and-shoulders above its other effects is the potential impact it will have on our jobs. While there are very different views about the long-term impact on employment as a whole, there is wide agreement that the development of Artificial Intelligence and robotics is set to have an enormous impact on the future of human work – driving up productivity, but in the process narrowing or completely shutting down many traditional jobs. So, the Pew Report's summary of "Key Findings" begins like this: "The vast majority of respondents to the 2014 Future of the Internet canvassing anticipate that robotics and artificial intelligence will permeate wide segments of daily life by 2025, with huge implications for a range of industries such as health care, transport and logistics, customer service, and home maintenance."

Our leaders need to ask searching questions, and not be put off by people trying to shut down the conversation because they pretend to know the answers. Because if we really don't know what lies ahead, the prudent leader will look at the risks and take damaging possibilities seriously. Even expert commentators on such issues sometimes change their minds, and policymakers need to be prepared for a shifting landscape. But how are they to assess the risk when both technology and opinion are in flux?

We've already referred to former Treasury Secretary Larry Summers, one of the few public figures who has actually addressed the issue – and changed his mind in the process. In a 2013 lecture titled "Economic Possibilities for our Children," which deliberately echoes the famous Keynes essay quoted as an epigraph to this book, he reflects on his first awareness of the question during his undergraduate days at MIT:

> There were two factions in those debates. There were the stupid Luddite people, who mostly were outside of economics departments, and there were the smart progressive people...The stupid people thought that automation was going to make all the jobs go away and there wasn't going to be any work to do. And the smart people understood that when more was produced, there would be more income and therefore there would

be more demand. It wasn't possible that all the jobs would go away, so automation was a blessing. I was taught that the smart people were right.

He goes on to say that he has had reason to change his mind and depart from the conventional wisdom: "Until a few years ago, I didn't think this was a very complicated subject; the Luddites were wrong and the believers in technology and technological progress were right. I'm not so completely certain now" (Summers, 2013).

Unless we are "completely certain," the threat needs to be taken more seriously. The scenario that Summers and other highly respected if outlier thinkers of left and right are laying out is alarming. The aim of this book is to encourage and inform the development of public debate and policies that respect the fact that they may be right.

Those who don't want us to worry often frame the discussion by referring back to what has happened in agriculture. Around 1870 something like three-quarters of Americans worked the land. They did this with the help of 25 million horses. Today, the number of Americans employed in agriculture is down to around 1 percent. But the result has not been massive unemployment. Plenty of new jobs eventually emerged for former farm laborers as the industrial economy took off; and as new opportunities became available, they acquired fresh skills. Not so for the horses. With the mechanization of agriculture, their labor was no longer needed. The new machines were designed to do – in Summers' phrase – "exactly what labor did before." The horses were fit only for the knacker's yard.

The contemporary situation has parallels with the Industrial Revolution of the eighteenth and early nineteenth centuries, as well as with subsequent tech-driven "revolutions," but there are also respects in which it is entirely unprecedented. As intelligent machines increasingly take over the human role in the workplace, the bottom-line question is whether, this time around, we are the farm laborers awaiting redeployment – or the horses. On any reading of the situation, our leaders need to prepare for a period of perhaps unprecedented turbulence in labor markets. The evidence is that they aren't.

Let's return to our two questions.

First, we have good reason to believe that, even assuming the conventional wisdom to be correct, *we are likely to face substantial turbulence as careers and industries are disrupted right across the economy before the hoped-for "new jobs" emerge in sufficient numbers to maintain the full-employment norm.*

Second, the possibility that this will not happen – that we shall instead see capital and technology incrementally substituting for human labor faster than new jobs can emerge – needs to be taken very seriously. It's a possible outcome that should be occupying our leaders; and our best thinkers should be addressing the question of how we might prepare. *There is a risk of the collapse of the "full-employment" norm* to which all the developed economies have become used. It may be hard to estimate how great that risk is, but it is not trivial.

In the chapters that follow we will review the advance of Machine Intelligence into the workplace, the arguments pro and con, the potential implications and mitigations, and the need to forge a consensus focused on risk that leads to action. And we will ask why this issue is not being taken as seriously as it warrants, either by experts or the public.

As things stand, policymakers don't know where to turn. The conventional wisdom, fed to them by their advisers, is that we need only double down on innovation and the jobs issue will solve itself. Meanwhile, a growing cohort of smart and distinguished individuals is beginning to see things very differently, and flagging what they see as a potentially major problem with fundamental implications for our social and economic assumptions about the future – and for policy.

Policymakers don't know which of these approaches is correct, and there is no mechanism available to them that will resolve the question. What they need is to synthesize these divergent possibilities into a single approach that is focused on risk.

That's my case.

ONE

Non-Human Resources

While experts disagree as to how many jobs will be taken over by machines, there is already a much wider range of applications of Machine Intelligence to tasks traditionally done by humans than many people realize. They include some highly skilled occupations which might have seemed immune to computerization, and the economics of this situation are potentially perilous for humans. As Norbert Wiener foresaw in the middle of the last century, the nearest parallel we have is the slave economy of ante-bellum America and the Greco-Roman world.

Let us remember that the automatic machine, whatever we think of any feelings it may have or may not have, is the precise economic equivalent of slave labor. Any labor which competes with slave labor must accept the economic conditions of slave labor. It is perfectly clear that this will produce an unemployment situation, in comparison with which the present recession and even the depression of the thirties will seem a pleasant joke.

Norbert Wiener (1950)

There have been several detailed studies assessing which jobs are liable to be taken over by machines in the foreseeable future. As Ron Shaich, CEO of the US coffee-house chain Panera, put it succinctly: "Labor is going to go down. And as digital utilization goes up – like the sun comes up in the morning – it is going to continue to go up." He's quoted in the World Economic Forum review, which then cites a Bloomberg report that robots could "replace up to half the US workforce within the next decade or two" (Peterson, 2015).

There are two ways of approaching this question – ask "experts" what they think will happen (as we saw in the Pew Report), or review the jobs and technologies one by one. The second approach has been taken by two Oxford economists, Carl Benedikt Frey and Michael A. Osborne (2013). Their painstaking analysis starts with the US Department of Labor division of the current labor force into no less than 903 detailed occupations. They whittle this down to 702 and proceed to assign values to each to assess their susceptibility to machine takeover. Starting with the jobs least susceptible to what they term "computerisation," at the top of their list we find recreational therapists. At the foot – most susceptible of all – are telemarketers.

What's a recreational therapist?

Their criteria are plain. While developments in Artificial Intelligence and robotics are not focused simply on mechanical activities and already

> allow many non-routine tasks to be automated, occupations that involve complex perception and manipulation tasks, creative intelligence tasks, and social intelligence tasks are unlikely to be substituted by computer capital over the next decade or two...the low degree of social intelligence required by a dishwasher makes this occupation more susceptible to computerisation than a public relation specialist, for example.

They build their approach on previous work done assessing how susceptible jobs were to off-shoring. Their conclusion is that about 47 percent of total US employment lies in their "high-risk" category: "jobs we expect could be automated relatively soon, perhaps over the next decade or two." What's more (and this assessment has been less widely reported in the press), a further 19 percent are at "medium risk," putting a total of 66 percent of US employment in the two categories.

Frey and Osborne's model predicts that "most workers in transportation and logistics occupations, together with the bulk of office and administrative support workers, and labour in production companies, are at risk." And: "More surprisingly, we find that a substantial share of employment in service occupations, where most US job growth has occurred over the past decades, are highly susceptible to computerisation." They ground their analysis in an interesting historical discussion of what's at stake, and cite not just the Luddite riots and similar popular efforts at resistance, but the famous example of Queen Elizabeth I's refusal of a patent that she considered would impoverish her subjects by taking their jobs: "It would assuredly bring to them ruin by depriving them of employment, thus making them beggars."

Their approach has been further developed, and also challenged, by a report from the Organization for Economic Co-operation and Development (OECD), the "think tank" of the leading industrial economies. While accepting much of the Frey-Osborne analysis, the OECD document (Arntz, Gregory & Zierahn, 2016) focuses on their assumption that the *tasks* they see shifting to Machine Intelligence (or "automatability," in the OECD term) essentially comprise the *jobs* they list. The OECD report sees more to many of these jobs than the particular tasks they involve. It also points out ethical and legal factors that will inhibit the machine takeover of many of these occupations. They conclude that 9 rather than 47 percent of jobs are on the line.

In later chapters we assess the relative merits of these approaches, and the question of whether new jobs will emerge as old jobs are taken over. But to begin let's sample some key fields and occupations where serious work is already going on to effect the switch from human to machine labor. In some cases we are discussing technologies already deployed. Others are just around the corner. But none of these prospects is merely a bright idea. They provide stark illustrations of the reason for the widespread "expert" uncertainty revealed by the Pew research and the detailed job characterizations of the Frey-Osborne and OECD reports.

We begin with the best-known and most striking example, that of self-driving cars or "autonomous vehicles." Then we sample more briefly a range of other employment areas. Finally, we look ahead nervously to examples of an emerging threat to what seem intuitively to be more "human-centered" and therefore "safer" occupations.

Self-driving cars

While Uber has seized the initiative in commercial vehicle deployment, it's Google that has captured the world's imagination with its development of the self-driving vehicles now motoring the public roads of California. What's less well known is that in parallel with Google every major auto company in the world is working on the same front, starting with what are being called "connected cars," including sophisticated electronics that help the driver drive the vehicle and connect with the company's base (CB Insights, 2016). Some of these models are transitional forms: they are essentially experimental, and will become, perhaps rather soon, obsolete. Others are capable of successive software, and perhaps hardware, updates on their journey to full automation. Few of us are aware of the degree to which cars have already become computers on wheels. The Ford F150 pickup, for example, hardly the most glamorous or seemingly high-tech vehicle, comes complete with 150 million lines of code (Saracco, 2016).

They also build on existing systems that we have long been used to, of which ABS (antilock) brakes may be the most interesting. The ABS essentially overrides the driver's instructions in difficult conditions and makes its own decisions about braking. Our cars have already begun to drive themselves.

One of the big uncertainties hanging over the discussion is whether the future of the self-driving car lies with Detroit or with Silicon Valley and the West Coast technology community. Judging by the clunkiness that has characterized much of Detroit's approach to the application of digital equipment to their current vehicles, the confidence of the traditional automakers that they know their business and the techies from the Valley don't may be misplaced. A case in point: I was recently driving a Cadillac SUV rental, with a visual display that told me my speed three times – the standard, and to my mind superior, speedo, with two separate digital displays saying the same thing. Add a further display with another large, colorful font telling me the percentage of oil remaining (it was 74 percent), and you begin to see the point. Then of course there was a center-column screen. The rear-facing reversing camera was a plus (perhaps the only obvious one from the digital get-up), though whether it outweighed the use of the same screen to set the heating controls (far more difficult when driving than traditional knobs, if not downright dangerous) is a good question.

It's certainly Google cars that have our attention, though also on the West Coast Apple is rumored to be developing a project, and Tesla with their innovative, high-end electric cars have been making bold claims. The Tesla cars can already drive in semi-automated mode on freeways – drivers are meant to be alert and ready to take over – and by 2018 a software update will enable them to drive across the country, coast to coast, in driverless mode. Among the European auto companies, Volvo has claimed that by 2020 their vehicles will be "accident-free," with the ambitious goal of ensuring no injuries to their users. Given Volvo's traditional reputation for safety, this is an interesting way to focus the development – not risking but enhancing what has made them famous. Scania, in which Volkswagen has a major stake, is working on a concept called "platooning" in which a series of trucks traveling in close convoy are controlled like a train from the lead vehicle. The Indian giant Tata has demonstrated self-parking technology, while Toyota is investing in research centers at MIT and Stanford (Hoge, 2016). *What*

Uber, meanwhile, is taking the lead in preparing the (largely unwitting) market for driver-free cabs. In a notable coup, they recently hired a labful of Artificial Intelligence researchers (forty it was reported) from Carnegie Mellon University to help them, and as we have already noted are rolling out a service in Pittsburgh as of this writing. Clearly they are planning to have their business model in place when the entire transportation economy begins to be overturned by the anticipated shift to a service- rather than ownership-based auto-usage model. We click on the Uber app, and in a short time a driverless Uber cab will appear. There has been some interesting discussion as to whether they will be able to roll their commanding brand presence in the growth of ride-sharing into the anticipated new situation where the market will have been enormously increased. They may well find that renting a driverless car has become a commodity, so the brand is irrelevant. In any event, their human drivers are likely to find their new careers cut short. Since Uber may be the only Silicon Valley start-up so far to have actually created large numbers of new jobs, the ending of their drivers' careers will be one of the more dramatic early impacts of the coming disruption (Samani, 2015).

According to *Wired* magazine, the shift to automated vehicles may begin with trucks, since their cost basis is so much higher, and automation would both enable 24-hour operation and slash insurance expenses (Davies, 2015). According to the American Trucking Association, there are currently 3.5 million truck-driving jobs in the US. Trucks cause many accidents – in 2012, 330,000 of them, killing 4,000 people (mostly, of course, in cars). Ninety percent of these accidents were the fault of drivers. Already, mining companies make extensive use of self-driving trucks (and trains), driven partly by labor costs but also by concerns about safety (Santens, 2015; Strauss, 2015).

The insurance point is an important one. It may seem counter-intuitive, but a major argument in favor of self-driving technology is safety. While we find the notion of 70 mph freeway traffic without a driver's hands on the wheel a scary prospect, the fact is that we humans are not really very good at many of the things we do (and this applies in many other areas where Machine Intelligence is set to take over). That's certainly why there are so many road accidents, with tens of thousands of fatalities a year in the United States and hundreds of thousands of injuries. Our powers of concentration are limited, as we know from the experience of sitting behind the wheel and suddenly realizing we have driven twenty miles without noticing. In fact 80 percent of the 35,000 Americans who die every year in road accidents are the victims of drunk driving, speeding, or distracted drivers – none of which is likely to be a problem for a computer. The prediction is that insurance rates will plummet, but will get more expensive if you want the luxury of being able to drive for yourself (though that may become a thing of the past sooner rather than later on freeways). Moreover, the prospect is not simply that vehicles will drive more safely than humans currently do, but that they will also drive defensively. With 360-degree awareness at all times, and communication with other vehicles (around corners, for example), they should protect passengers from the mistakes of other drivers as well.

One of the beneficial effects will be a likely decline in the need for Emergency Room admissions (with a knock-on significance for healthcare costs). Of course, there will then also be implications for healthcare jobs – as well as for jobs in a slimmed-down auto insurance industry. Meanwhile, a standard liability objection to the autonomous vehicle scenario has already been addressed by Volvo. Who, people ask, will be responsible if an accident occurs when the car is driving itself? Volvo has taken the lead in stating that it will accept full liability – a policy that may set the standard for other manufacturers (Insurance Times, 2015).

A report from Lloyd's of London (2014), the insurance underwriter, has already addressed some of the practical issues raised by the prospect of self-driving cars. The report visualizes an economy in which private ownership has declined radically and much greater use is made of self-driving fleet vehicles, perhaps rented by companies from their manufacturers rather than owned. Lloyd's recognizes some of the problems in such a scenario:

> With less reliance on a human driver's input, however, increased risk would be associated with the car technology itself. Computers can do many things that a human driver cannot: they can see in fog and the dark,

and are not susceptible to fatigue and distraction. However, they can also fail, and systems are only as good as their designers and programmers. With an increased complexity of hardware and software used in cars, there will also be more that can go wrong.

As a result, they expect that the human in charge will still need to be ready to take over. "If drivers are expected by law to supervise an autonomously operating car, they may find it difficult to remain focused on this task if they feel able to trust the car."

The scenario envisaged by Lloyd's is a transitional one, with "self-drive" operating like cruise control as a driver option. By contrast, Google's latest generation of prototypes were intended to have no steering wheels at all. They were added at the insistence of the State of California. In fact, such a half-way house situation is more problematic than the full-on driverless scenario with no fall-back controls. Keeping *half-focused* on the driving process – for human drivers who tend not to focus properly even when they are supposed to be driving entirely on their own – seems particularly dangerous.

What's more, self-driving technology is likely to change the market for car ownership in dramatic fashion. If clicking on an app brings a car to your door almost immediately, why would you wish to own one? In cities there is already some evidence that the Uber car-sharing service is diminishing vehicle ownership as it is so much more convenient (and often cheaper) than calling a cab. Car-buying among Millennials has been falling for some time (Schwartz, 2015). Indeed, a University of Michigan study has argued that vehicle ownership could be cut by as much as 43 percent. The study notes that currently the average US household owns 2.1 cars; it suggests that this could drop to 1.2 (Schoettle & Sivak, 2015).

This may in fact turn out to be an underestimate. The researchers assume that the typical American household will still end up owning a car – just getting rid of car number two since driverless rentals are convenient and cheap. It's more likely that once the revolution starts it will keep going. Another estimate of future car ownership envisages a 60–90 percent reduction, and draws out the dramatic impact that is likely to have on the auto manufacturing industry. If we don't need two cars, why do we need one? The cost implications for consumers could also be dramatic. In *Humans Need Not Apply*, Silicon Valley entrepreneur and Stanford academic Jerry Kaplan (2015) lays out some interesting math. Starting with numbers from the American Automobile Association (AAA), he suggests that the cost of personal travel

What about companies? (yes.)

could drop by 75 percent, from 60 cents a mile to a mere 15 – freeing up a big slice of the average family's disposable income for other purposes, and, according to another study, benefiting the US economy by no less than $3 trillion a year. Of course, we have heard similar optimistic predictions before. In the 1950s it was claimed that nuclear energy would give us free power. But the closer we come to the actual adoption of self-driving technology the better these estimates seem to be.

The half-way house Lloyd's report reveals the same unease about our moving beyond vehicle ownership. Lloyd's commissioned a UK survey that found most people were nervous about abandoning driving. "People are uncomfortable placing their trust in a computer and they like the process of driving too much to give it up." This is of course how people tend to think before they get used to a dramatic innovation. It's hard to see them still thinking this way once the service is available, plainly safe, convenient – and economical.

The economical point is a basic one. While cars are extremely expensive (for most of us the most costly thing we buy after a house), they spend most of their time in the garage or sitting in the street. As *Business Insider* journalist Matthew DeBord (2015) reports, one Wall Street analyst reckons that the typical utilization rate of a car is 4 percent. It is used for just one hour a day. And "this has not changed materially for a hundred years…To him, this is a stunning waste and an unsustainable model." Of course, this argument is up against the convenience of ownership: the car is there when needed, waiting to be driven. "I have a car that spends its entire day in my driveway," says DeBord. "But I really, really like that my car is there when I need it."

While rural areas may long retain a different set of practices, as they do in other respects, the smart betting is on cities and suburbs succumbing to the driverless/rental vision. And Uber's existing service is helping mark the way. Already, many of us feel differently about renting an Uber than we do about calling a cab. And at the Frankfurt auto show, luxury car-maker Mercedes-Benz recently announced its own plans for autonomous cabs: a "free-floating car sharing provider" that "lets customers quickly hire vehicles left dotted around cities." "It would be even more convenient if the car came to you autonomously," said Dieter Zetsche, boss of the parent company Daimler. "And it would be extremely practical if [it] appeared without needing to be prompted, once my appointment in the calendar had come to an end" (Vincent, 2015).

The wider social implications are enormous. Think of the freedom of movement that will come to those who currently cannot drive – whether because they are sick, or elderly, or disabled, or indeed young. Think of being able to work – or party – on your trip (yes, alcohol restrictions will no longer make any more sense than bans on texting and phone use). It's been suggested that we will take more road trips than in the past, as door-to-door service in a work (or sleep) environment may make air travel for distances of only a few hundred miles less appealing. And the sleep environment may cut hotel usage, or lead to a new kind of motel designed to partner with autonomous vehicle use (after sleeping in the car en route, we need to shower on arrival). And this is just the beginning – the result of applying Machine Intelligence to one simple human activity, driving.

Travel.

Legal, financial, computer services and management

As driverless cars show, the impacts of Machine Intelligence can go much further than first meets the eye. While plainly mechanical applications of Machine Intelligence such as ATMs and supermarket checkouts make most sense to us, the reach of the robotics/AI revolution extends well beyond such low skill-set tasks.

"Sophisticated algorithms are gradually taking on a number of tasks performed by paralegals, contract and patent lawyers. More specifically, law firms can now reply on computers that can scan thousands of legal briefs and precedents to assist in pre-trial research." As an example, one system proved able to analyze and sort 570,000 documents within two days (Frey & Osborne, 2013).

The point is that the computer role here does not lie simply in searching text and databases but increasingly includes altogether more subtle tasks. This should surely worry those who have too quickly assumed that the employment impact will lie at the lower end of the skill and economic scales. Occupations that require subtle judgments are also increasingly susceptible to computerization. In relation to many such tasks, the unbiased decision of an algorithm represents a comparative advantage over human operators. In the most challenging or critical applications, as in ICUs, algorithmic recommendations may serve as inputs to human operators; in other circumstances, algorithms will themselves be responsible for the appropriate decision-making.

Perhaps most strikingly, "Even the work of software engineers may soon largely be computerisable...Big databases of code also offer the eventual prospect of algorithms that learn how to write programs to satisfy specifications provided by a human... Such algorithmic improvements over human judgment are likely to become increasingly common" (Frey & Osborne, 2013). Frey and Osborne proceed to quote one estimate that "sophisticated algorithms could substitute for approximately 140 million full-time knowledge workers worldwide." This is dramatic, indeed rather chilling, news – and highly relevant to the push for technical education ("STEM") as the solution proposed by conventional wisdom to the challenges of job losses through automation.

Moving higher still up the professional hierarchy, listen to Lucy Marcus (2015), global guru on boards and company leadership: "It might sound far-fetched, but 45 percent of 800 executives surveyed by the World Economic Forum's Global Agenda Council... said they expected an artificial intelligence machine will sit on a company's board of directors by the year 2025." She goes on to say that she is with the 55 percent, and skeptical of the notion that AIs will actually sit on boards; data will become yet more important to decision-makers, but the decisions will remain with (voting) humans. She may be right, in the short-to-medium term. But it is a mistake to see Machine Intelligence as merely offering raw data to people who then need to make the decisions – that is, once those decisions get complicated.

As has already happened in the financial markets (and proved an accelerant in the 2008 collapse), machines are increasingly being "trusted" with complex decisions that defy human grasp. Whether or not this will entail "board membership" in standard terms (as in, participation in an elected slate), there will be a growing reliance – perhaps a shareholder insistence – on the use of and in some cases deferral to Machine Intelligence in high-level corporate decision-making.

Education: the MOOC phenomenon

What about highly skilled professions such as teaching? Several years ago there was a flush of interest in what we used to call online courses. They were re-named MOOCs ("massive, open, online courses"), a trendy acronym, as advances in the deployment of Artificial Intelligence enabled a new generation of online educational offerings. Their magical key ingredient is zero marginal cost. That is, one extra student does not cost any more money. One reason traditional online education has not made the strides many expected twenty-five years ago is that it is not far removed from the old mail-based "correspondence courses" that date back to the nineteenth century. Standard online courses have a special convenience – especially where they are asynchronous, that is, you can study any time – but they can involve as intensive an effort for their teachers as does old-fashioned classroom education. So, for example, classes need to be small. What has changed with the advent of MOOCs is the deployment of courses that cost no more to teach a million students than they do to teach a dozen.

When a technology catches the headlines but fails to catch on in the marketplace it does not take long for some commentators to write its obituary, and the MOOC phenomenon falls into this category. But it is very early days. Major institutions – including top universities – are investing, if somewhat warily, in a technology that could up-end the entirety of higher education. And, as with driverless cars, we need to note the economic factor here. Not only is traditional college education becoming increasingly expensive for both students and governments, but this fact (and allied concerns about student debt) is now a major talking-point. When the possibilities offered by a new technology meet an economic opportunity rapid change can be expected to follow. While the enthusiasm that led the *New York Times* to call 2012 "the year of the MOOC" has quietened, it is notable that innovation guru and Harvard Business School professor Clay Christensen has made a startling longer-term prediction: that within fifteen years, half of US universities and colleges may be bankrupt (Suster, 2013).

The early experience of the several platforms offering these courses (both for-profit and not-for-profit) has been mixed (Wharton, 2015), which is why the initial hype has given way to a more sober approach to their development and prospects. It's understandable that the major universities which have dipped their toes in the water have done so with a degree of trepidation, since they have no wish to undermine their own market for students and degrees. Hence a focus on offering non-credit courses, in some cases with "completion certificates" (generally for a fee) for those who wish to claim their achievement for professional purposes. Somewhat surprisingly, we have yet to see an offer of a full undergraduate online degree program, though this may simply be a matter of time. The first full grad program has already been tested, by Georgia Tech in collaboration with MOOC provider Udacity and AT&T – the first cohort consisting of company employees. What's interesting about this offering is that it is not a watered-down qualification but the equivalent of the on-campus program. This will be key to the market for such courses, though it is plainly costly to initiate such an effort and will involve intense competition –

especially if the "branding" is that of a top institution (Zhenghao et al., 2015).

Whether or not Christensen's apocalyptic scenario of the collapse of half of US higher education proves justified, the MOOC principle of zero marginal cost offerings that in time have full credit stands to undermine the current system fundamentally. And, in the process, it could eradicate tens of thousands of jobs for college professors. This is well illustrated by a recent Schumpeter column in *The Economist* with the arresting title, "Professor Dr Robot QC" (Schumpeter, 2015). And the subtitle runs: "Once regarded as safe havens, the professions are now in the eye of the storm." After citing a book from the 1930s that described the professions as "centres of resistance to crude forces which threaten steady and peaceful evolution," the writer turns to new claims that the core strengths of the professions are under threat from information technology. Already, algorithms "allow the laity to dispense with some professionals entirely, or at the very least to take them down from their pedestals. Every month 190 million visit WebMD – more than visit regular doctors in America." Even visiting the doctor increasingly involves Machine Intelligence: IBM's Jeopardy-winning computer Watson is now being offered as a medical brain to partner hospitals. As Frey and Osborne

(2013) note in their report on the employment impact of Machine Intelligence, oncologists at the Memorial Sloan Kettering Cancer Center are already using Watson "to provide chronic care and cancer treatment diagnostics." By 2013, Watson had already absorbed 600,000 medical reports, 1.5 million patient records and clinical trials, and 2 million pages of text from medical journals. It should be no surprise that an early application was in the area of cancer care, since this is enormously expensive and particularly benefits from the computer's ability to compare patients' individual symptoms, genetics, family and medication history, for diagnosis and treatment plans.

Returning to *The Economist* article: "Educational apps are the second-most popular category in Apple's app store after games, and MOOCs are attracting millions of students. Judges and lawyers are increasingly resolving small claims through 'e-adjudication'. It is one of the techniques employed by eBay to settle the more than 60 million disagreements among its users each year" (Schumpeter, 2015). How far will this go? Schumpeter asks. "Messrs Susskind and Susskind predict that it will go all the way to 'a dismantling of the traditional professions'. These jobs, they argue, are a solution to the problem that ordinary people have 'limited understanding' of specific areas of expertise.

But technology is making it easier for them to get the understanding they need when they need it." The Susskind book offers a sophisticated reflection on both what constitute the "professions" in contemporary society, and the subtle ways in which much professional work is potentially susceptible to machine alternatives (Susskind & Susskind, 2015).

Elder care and nursing

The Machine Intelligence revolution clearly challenges assumptions that only low-skilled jobs are at risk. But what about those roles with an intrinsically high (and irreplaceable) "human quotient"?

One of the most striking developments here lies in the care of the elderly and sick. It's plain that as our societies age, the task of providing such care will grow substantially. This is especially true of certain countries whose demographics are more skewed than others. It should therefore be no surprise that much of the focus in developing elder-care applications has been in Japan, where the shifting demographics have made this a major issue. According to one report, already in 2010 Japan had an astonishing 30 million individuals in care facilities (Hay, 2015).

old PA don't want feel the you

But the economics of elder care are universal. One estimate suggests a population increase for the over 65s of 181 percent by 2050 – a stunning difference from the projected increase of those between 15 and 65 of only 33 percent. So both the UK and the US have seen substantial engagement in the development of technologies to make it simpler to care for those who are dependent. At a basic level this may just involve sensors in the homes of those living alone which will report if their pattern of movement changes and trigger a visit from a relative or nurse. Other initiatives include the robotic companions Paro – a mechanical seal that responds to petting and will cry if dropped or ignored (an update for the elderly of the kids' Tamagotchi of a generation ago) – and Palro, which can "play games and dance with the elderly, keeping their minds active with trivia" (Hay, 2015). The approach can be very simple: take the Roomba automated vacuum cleaner, or a robotic arm devised to help people with limited movement to shower, or the general purpose robots being developed to wash dishes and deal with laundry. While none of these is focused on strictly medical needs, they variously address the care of people whose situation is more fragile than it was when they were younger.

At a much more sophisticated level, we have robotic carers and companions. ChihiraAico is designed to look like a thirty-two-year-old Japanese woman and make people more comfortable talking to a robot about their problems. "And just earlier this year, SoftBank released Pepper, one of many personal robots, a humanoid creature with the power to read and respond to human emotions. A massive breakthrough in our ability to connect with robots, Pepper could do wonders for the mental engagement and continual monitoring of those in need" (Hay, 2015).

A key issue for ethics and social acceptability will lie in the degree to which we (as relatives and friends – and as potential patients) feel comfortable with robots behaving like people, and with robots being assumed to be people by patients: "Robot caregivers can listen to even the wildest story or one that is nonsensical without dampening the spirits of the person with dementia" (Birkett, 2014). As natural-language recognition and use become increasingly sophisticated, we may expect this set of questions to become yet more acute.

Aside from the potentially controversial questions that some of these applications raise, we are facing the evolution of Machine Intelligence capacities that are replacing many dimensions of the roles traditionally fulfilled by human carers (those which would previously have been considered too difficult to computerize) – whether family members or professionals in healthcare and social work. As we have noted above, when there is a strong economic argument at work we may expect rising interest in developing and using such capacities.

Psychology and psychiatry

If you are planning for a career that will survive competition from machines it might seem an obvious choice to focus on the exploration of the human mind. Psychologists come in at number 17 in the Frey-Osborne taxonomy – with pretty much as high a "human quotient" as can be imagined. Yet as I write this chapter my Twitter feed just turned up another article from *Scientific American* about advances expected within the next year in the field of Artificial Intelligence and emotion:

Here at Carnegie Mellon University, in our Robotics Institute, Fernando De la Torre has led development of some particularly powerful facial image analysis software, called IntraFace. His team has used machine learning approaches to teach IntraFace how to identify and track facial features in a way that is generalizable to most faces. They then created a personalization algorithm that enables the software to perform expression analysis on individuals. It's not just accurate, but efficient; the software can even run on a smartphone.

De la Torre and Jeffrey Cohn, a psychologist at the University of Pittsburgh, already have had encouraging results in detecting depression in psychiatric clinical trials. Detecting depression "in the wild" requires the ability to capture subtle facial expressions, which they are doing.

Imagine the possibilities: A virtual psychiatrist could help diagnose depression by analyzing the emotions we display during clinical interviews; it could even quantify changes in mood as the disease progresses or as therapies kick in. Marketers could better gauge how audiences respond to their products and ads, while teachers could assess whether a lesson plan was fully engaging students. Smartphones might alter directions and advice if they perceive that we are upset or confused.

In other words, our passionless devices will come to know us through the emotions we all wear on our sleeves. (Moore, 2015)

These developments should be seen in light of MIT professor Rosalind Picard's pioneering work in what she has named "affective computing," which suggests that it may be naive to believe that the "human quotient" line is impenetrable. Plainly, the skills we would most naturally associate with the need for human presence and intuition are capable of being broken down into digitizable form.

Or, in other words, all bets are off. As we shall see when we come to assess these developments later, our expectations about which jobs are "safe," and thus the mental model we bring to assessing potential future developments in labor markets, have constantly been upset by events. This is a problem for all of us, but especially for those who keep seeking to peddle reassurance. If AI is moving to occupy the space of emotional intelligence, then the workless world just became a lot more likely.

The threat is not just to the jobs of humans – or horses. For the 1988 French film *The Bear*, more than fifty trained bears were auditioned. But in the Oscar-winning blockbuster of 2015, *The Revenant*, this was not necessary, even though the story centers on a protracted fight between the lead character, a trapper played by Leonardo DiCaprio, and a grizzly. No bears auditioned; not a single one was needed. The bear's part was created entirely by digitization (Hawkes, 2016). Of course, while bear wranglers could see the writing on the wall, the AI approach ensured plenty of work for the digitizers.

TWO

"The Stupid Luddite People"

While the future is unknown, intelligently assessing what is likely to happen is key to our current decision-making. One major reason people are reluctant to address the potential impact of technology on employment is that they don't want to be labeled "Luddites." It's an easy accusation to make, and threatens to close down the conversation before it gets going. We noted in the Introduction that one of the handful of public intellectuals questioning the conventional wisdom is former US Treasury Secretary and Harvard President Larry Summers. Since he tackles the "Luddite" issue head-on, his words offer a perfect starting-point for this next chapter.

When I was an MIT undergraduate in the early 1970s, a young economics student was exposed to the debate about automation. There were two factions in those debates. There were the stupid Luddite people, who mostly were outside of economics departments, and there were the smart progressive people... The stupid people thought that automation was going to make all

the jobs go away and there wasn't going to be any work
to do.

Larry Summers (2013)

There's growing sensitivity on the part of some technology boosters to criticism – any criticism – of where
this is all leading. Yet as we have learned from every
disaster in history – from the sinking of the *Titanic*
to the GMO fiasco to the 2008 Wall Street collapse –
discouraging critical thinking always raises risks. The
bigger the issue, the greater the risks. Even left-of-center
politicians and labor leaders have been reluctant to get
drawn into this discussion, for fear of being labeled
Luddites. The techno-boosters are making it difficult
to talk about something that desperately needs to be
discussed. By doing so, they may not actually be the
friends of technology they think they are – in the long
term. Anticipating potential problems early on is key
to ensuring success, and crucial to that is giving space
to dissonant voices. From thalidomide to "mad cow"
disease, from smoking to "genetically modified" organisms, we have seen that disparaging early critics only
leads to more damaging consequences.

Yet there are those who wish to suppress dissonant
voices where they raise hard questions. The most striking example is the bizarre suggestion that Elon Musk,

[margin note: they own the communication channels too]

whose innovation empire includes Tesla cars, SpaceX rockets, and the proposed hyperloop ultra-fast transportation system, is not merely a Luddite, but qualifies as "Luddite of the Year." This was the judgment of an esteemed Washington DC think tank.[1] Why? Because Musk dared to sign a statement from some of the world's top scientific thinkers (including Stephen Hawking) that raised questions about the long-term dangers of Artificial Intelligence essentially taking over from humans. Raise a potential problem about technology, and you are in danger of someone calling you a Luddite. It's hard to imagine anything that would serve as a better inhibitor of full and free public discussion of the risks entailed in emerging technologies. Yet this is the time for *Homo sapiens* to encourage the frankest and most honest conversation about what lies ahead.

Moreover, the impact of robotics on employment is a subset of a broader discussion about the extent of the impact of the development and application of digital technology. And the techno-boosters who dismiss concerns about technological unemployment oddly find themselves on both sides of that debate. To illustrate: I was recently invited to take part in a conference in Madrid, Spain, hosted by *The Economist* magazine. The theme of the conference is explained as follows:

IMPACT OF THE DIGITAL AGE: ARE THESE REVOLUTIONARY TIMES FOR THE GLOBAL ECONOMY?

Some claim that we are now entering a second machine age the effects of which are yet to come. Others paint a much more sombre picture doubting the transformational capacity of digital technology and the pace of innovation. Undoubtedly innovation is vital to improve efficiency and increase productivity....the late 19th and early 20th centuries produced technologies with more far-reaching benefits to our societies than anything IT and digital evolution has achieved so far.

The two sides of the debate convene for a discussion on what digital progress means for the world economy. Does recent innovation amount to a digital revolution? Or has innovation and new technology slowed down and stopped driving increased productivity and economic growth? Have we reached a technological plateau? Moreover, revolutions have beginnings and endings. If this is only the beginning of a digital revolution then where is the end? When will we feel the full impact of the innovations we are struggling to adopt now?

It was interesting that *The Economist* captured the issue in these terms: "Are these revolutionary times for the global economy or not?" Clearly, how we answer that question will shape how we tackle the jobs issue. Is this a real revolution, rather than just another ultimately routine uptick in the standard progression of technology and the economy? The techno-boosters want to answer Yes, and yet to deny that it could have a dramatic impact on employment. It's as if jobs had been made a "safe haven" from fundamental disruption. Absolutely everything is going to change, they say – except that we will still have jobs for pretty much everyone like we had in the old industrial economy. This is certainly a very comforting assumption, but it is also truly naive.

The mythical Ned Ludd wasn't much interested in a serious conversation; he preferred the use of the sledgehammer. So it's ironic that it's now the techno-boosters who feel the need to wield such a weapon of their own in order to avoid a serious discussion. We need to stop calling people "Luddites" and start a respectful roundtable conversation that is inclusive of those who are beginning to challenge the conventional wisdom.

The most influential recent book on the issue has been MIT economists' Erik Brynjolfsson and Andrew McAfee's *The Second Machine Age*, subtitled *Work, Progress, and Prosperity in a Time of Brilliant Technologies* (2014). It's based on their earlier e-book *Race Against the Machine* (2011). They set out in detail the case for the transformative impact of Machine Intelligence on employment, and conclude both that it does indeed represent a fundamental challenge to the labor market, and that there are policy and other initiatives that can mitigate its impact. Entrepreneur and writer Martin Ford's *Rise of the Robots* makes a similar case, though more trenchantly. More nuanced contemporary analyses, which also address the question of current wage stagnation, include James Bessen's *Learning by Doing: The Real Connection Between Innovation, Wages, and Wealth* (2015).

A key article offering a strong caution to the view that technology automatically destroys jobs comes from another MIT economist, David Autor. In "Why Are There Still So Many Jobs? The History and Future of Workplace Automation" (2015), he reflects on the impact of recent technologies on labor markets and offers principles that give a more optimistic view of what might come next. Correctly assessing the impact of technology on employment "requires thinking about the range of tasks involved in jobs, and how human labor can often complement new technology." The former point is elaborated by the OECD report we introduced in Chapter 1 as one reason for its assessment being more optimistic than that of the Oxford Martin School's approach – because "jobs are made up of many tasks," technological substitution is more complex than the first assessments might suggest. Moreover, Autor draws on the work of the philosopher Michael Polanyi to suggest that there is an element of "tacit knowledge" in much human action; that is, we do things that we ourselves imperfectly understand – such as writing a persuasive paragraph, breaking an egg on the side of a bowl, or identifying a rare bird species from a glance. Hence among those tasks "most difficult to automate" – both at lower and higher skill levels – are those involving "flexibility, judgment, and common sense," components that automation has found and likely will find most challenging to emulate. Autor also focuses on what will be the increasingly crucial question of the complementarity of humans and machines in enabling successful job performance and the creation of value.

It's plain that any serious analysis needs to take such considerations into account, although one might be forgiven for pointing out that until recently one very solid example of the role of "tacit knowledge" in lower-level human performance – one familiar to all of us – would have been that of driving. It has not taken long for the tacit knowledge that leads good human drivers into a succession of intuitively complex choices to be turned into algorithms. And in terms of the general application of this principle at a higher level, the recent work of Susskind and Susskind on the implications of Machine Intelligence for the professions – already discussed – is as alarming as it is fascinating.

Things really can change

So, how do we engage in serious discussion? We can start by taking a more informed approach to thinking about "the future." As we look back, we just keep being reminded of how things change. It's hard to imagine life before our cellphones gave us driving directions, before Skype made phone calls free, Facebook kept us effortlessly in touch with our friends and families, and Google gave us access to the world's store of knowledge. All this has taken place in just two decades.

Back in 1995, when only 18 million Americans were online, there were none of these, but some startling things had begun to happen. Craigslist was launched. And eBay. And Amazon. And the first dating site, Match.com. It was also the year of the Netscape IPO, and Microsoft's launch of Internet Explorer. All in one year.

Each of these changes has been powered by computer chips, operating according to what we have come to know as Moore's Law, the "law" proposed by Intel founder Gordon Moore that has held true for half a century. It states (in simplified form) that computer processing power doubles every two years. In other words, anything driven by computer technology – anything that can be digitized – is advancing at an exponential pace. It gets faster, much faster, every year. This makes thinking about the future even more difficult. But it also makes it even more important. The startling shifts in our lives since Google was founded in 1998 will be dramatically overtaken by the changes that come next – whatever they may be.

It's worth reflecting on the significance of the exponential idea. We are familiar with it in finance – compounding interest is what drives long-term returns to our retirement investments. But its power is best illustrated by the old tale of the grains of rice and the chessboard. Here's one version: An Indian prince challenged a visiting wise man to a game of chess, and offered him anything he'd like to name as a prize. The wise man said all he wanted was a few grains of rice. One on the first square, two on the second, four on the third, then double amounts square by square. The math is simple enough if we don't get put off by large numbers. By the twentieth square, one million grains of rice are needed. By the fortieth, the number has risen to a billion. And there are still twenty-four squares to go.

By the final square, the amount of rice needed comes to 18,000,000,000,000,000,000 grains, which has been calculated to weigh around 210 billion tons (depending on the kind of rice).

While there is evidence that Moore's Law may at last be slowing down (Technology Quarter, 2016), the prospects for quantum and biological computing suggest huge changes ahead. And there are graphic illustrations of how far we have come. It's been calculated that the computers that took the *Apollo* astronauts to the Moon had just twice the power of a Nintendo. The iPhone 4's computer is comparable to the Cray-2 computer, the fastest machine in the world in 1985. In fifty years, our computing power has increased by a factor of one trillion (Stone, 2015).

Plainly, the Exponential Principle makes it extraordinarily difficult to predict what comes next. But it does tell us to expect that the changes awaiting us will be huge. The chessboard illustration is especially telling. For the first few squares, the quantities of rice are growing, but quite manageably: 1, then 2, then 4, then 8, then 16, then 32, then 64. But by the time we get to the second half of the chessboard (a phrase that has been attributed to inventor and futurist Ray Kurzweil (2000)), the increase from just one square to

WHAT

the next has become enormous. Now we are a half-century into the computer age – and Moore's Law – the rate of change we must expect year to year is far greater than it was.

The fallacy of the new normal

Yet that isn't how we usually see things. We do know that change is happening fast, and that it's getting faster all the time. But how do we cope? One of the most influential books of the twentieth century was Alvin Toffler's *Future Shock*, which first appeared in 1970. The title came from the idea that people experience "culture shock" when they shift between different cultures. Toffler's point was that the future is like another culture. And moving into it fast would not be easy.

It may be amusing to reflect that, back then, when change was so much slower, people were concerned that we would find it hard to manage. But change is never easy. And while we are actually far better at adapting to it than we expected, what we find especially hard is looking forward and imagining the continuation of rapid, disruptive, disconcerting changes. We tend to look back and sigh with relief that we have made it through all the disruptions. So when we do look ahead, we expect tomorrow to be very like today.

Think of it as a graph we are traveling up. We look back and see how fast and steep the curve has risen to bring us to where we are now. And we look forward, and the line we see is much smoother. We put a kink in the curve, and flatten it. The line of the future keeps going up, but more gradually. We simply don't imagine that the dramatic pace of change that got us here can continue – let alone speed up. We've established a "new normal" as the baseline for tomorrow.

Yet as the image of the rice piling up on the chessboard illustrates, smoothing the curve ahead of us is actually the opposite of what we need to do if we truly want to think ahead. It is only going to get steeper. Steeper and steeper. And while there are aspects of our lives that have been largely unaffected by the digital revolution (cooking, sports, the weather), anything driven by computer power will be driven up that curve. Think back to the world in which many of us were raised – before maps became GPS, telegrams became texts, and encyclopedias became Google. Try sitting down and making a list of all the things you use your "phone" for, and the dozens of processes and activities that used to be needed to accomplish them, from buying CDs to spending an evening a month writing out checks. The changes ahead will be faster and greater, and we're in denial if we think otherwise.

We all *tend* to think that the future will be plain sailing. Developments in technology will make life easier and more fun, but without any special threat to what we now take for granted as normal. Of course, the challenge to labor markets isn't alone in posing a fundamental threat. Equally significant, and driven by the same tech forces, are the implications of the Internet of Things – every object in the world linked up – for safety and security, as hackers can break into our cars and homes and take control of them in the same way as they have already been able to steal our credit card information in the past. But we humans tend to want to focus on the good news, and that's plainly how our political leaders' minds work, left, right, and center. It's the kind of thinking that their electorates are likely to find appealing. The short-term character of democracy doesn't help — if your job is on the line in two years, or four, then why make yourself unpopular with scare stories about what lies twenty years out? Of course, the challenge of leadership is to frame issues in the longer term and prepare people's minds for change.

In every walk of life we meet the same tendency; even business leaders – and technology gurus – are not immune. One encouraging sign is that great business leaders – Jack Welch of GE is often seen as the best example – while accountable to daily stock price fluctuations, and quarterly earnings calls with analysts and investors, have nevertheless been able to build great companies by thinking decades ahead. That's the kind of far-sightedness now required of every global leader as we clamber up the ever-steeper stretches of the Moore's Law curve. The length of time it has taken global leaders to be responsive to scientists' predictions about climate change highlights how difficult this process can be. Their general failure to address the equally significant issue of the threat to the future of antibiotics offers a further example of the same problem.

Why is it then that so many opinion leaders ignore the potential impact of intelligent machines on jobs – and techno-boosters actually tell those of us raising the issue that we are stupid? The boosters believe they are the true guardians of technology and its vision. They are the "true believers" in its revolutionary character, and the problem-raisers are standing in the way of the future. They say we shall have no problem adapting to the impact of Machine Intelligence. We shall find new jobs. We shall flourish doing the things that machines find too difficult. They may be right, but in their zealous over-confidence they are buying into the Fallacy of the

New Normal. These same people are kinking the curve, imagining a future more comfortable than they actually have reason to believe it to be.

One understands the appeal of what we might refer to as technological determinism – the idea that technology development itself drives change and shapes the future of society, and that it will solve the problems it creates so we should leave well alone. But history is littered with examples of the interplay of humans, technology and economics in which human foresight, courage and planning made all the difference to what happened – from interventions in the face of stock market and banking failures (from Roosevelt's New Deal to the global response to Lehman Brothers' collapse in 2008) to direct technology regulation (asbestos, DDT, auto emissions, the whole environmental protection enterprise) to the public/political role in funding science and technology (it's well known that most of the features in Apple's iPhone and its Android equivalent were not the result of Silicon Valley vision but of US government investment, mainly through DARPA, the Defense Advanced Research Projects Agency – which, among other things, invented the internet itself). We must not lose sight of this; but nor should we assume it will take care of itself.

Thinking clearly about the future

So, how should we think about the future? We need to take it seriously, even if we are not sure which future it is likely to be. Let's look at all the plausible futures, decide how probable each one is, and do our best to prepare for all likely outcomes – and their respective risk profiles.

A simple approach to assessing risk starts by assigning possible outcomes to four categories – four boxes in a simple matrix: High Impact and High Probability; High Impact and Low Probability; Low Impact and High Probability; Low Impact and Low Probability. Plainly, the first two are the ones that matter most. And the second is the trickiest.

There are in fact two separate arguments here, and they can be distinguished for the purpose of risk analysis. One refers to the likelihood of serious structural unemployment; the other to the likelihood of a failure to recover from such turbulence in the labor market and the long-term whittling away of the economy's capacity to sustain "full employment" in a future in which AI/ robotics has been substantially deployed. Is there a High or Low Probability of serious structural unemployment? Either way, if it happens it will have a major impact on developed (and potentially developing) societies. If there is a failure thereafter to recover "full employment" levels – which is the scenario suggested by Gates, Summers and Murray – then the impact becomes very much greater.

As we face the future, it's difficult to see how policy-makers can avoid some hard thinking about the future of jobs.

THREE

Welcome to the Rust Belt

One of the big surprises of the twenty-first century is that leaders in government and industry, and even in labor unions, do not see the prospect of wide-scale labor market disruption as a serious possibility. It is not a theme of their speeches. Or their planning. As we watch political candidates debate, there is never a word about the complex set of policy challenges it could entail. Despite the fact that Bill Gates and other tech visionaries have begun to break the silence, there's little evidence of this having an impact on mainstream thinking. Yet we can't avoid asking tomorrow's questions if we are to answer today's. How likely is it that we can expect to face major disruption in our labor markets? As we've seen, after detailed analysis of those "high-end" occupations we call the professions, the Susskinds challenge the easy but potentially reckless assumption that there will always be tasks left for humans.

Increasingly capable machines, we conclude, will gradually take on more and more non-routine tasks; and so the intuition that there will always be tasks left that only humans can perform will prove to be ill-founded.

Richard Susskind and Daniel Susskind (2015)

Industry after industry is in the process of dramatic transformation, and in every case the result is the same – more convenience, lower prices (often, free services), and fewer jobs. But this process is just getting going. It would be naive to believe that the most dramatic transformations are behind us. We are just at the start. And at the heart of what is happening lies something both very simple and very profound: One at a time, we are transforming functions that humans used to perform into functions for machines.

At one level, this has always been the story of technology. The wheel, levers, pulleys, farming implements like the adze and the plow, have all enabled us for thousands of years to do things that would be much more difficult without them, and at the same time required fewer people to do the work. The point is that the pace is picking up – very fast. At a more fundamental level, we are steadily removing the labor factor from the creation of value. This is already obvious in some of the examples we have mentioned. While it once took a clerk with skill and experience to book an airline ticket, it's now something we do for ourselves using a cleverly programmed machine. And what we do is much more sophisticated than what the clerk could do. I just booked a ticket to visit my daughter in Seattle. In the process I discovered the cheapest days and times to fly across several airlines, reviewed a couple of dozen possible itineraries, decided whether to pay with miles or dollars, picked my preferred seats on five different airplanes, and punched in my credit card number. The point is: There was no other person involved. The computers had been programmed to help me sort through the data, and across at least four companies – from Travelocity to United Airlines to Visa to Citibank – they were all talking to each other. They did so across the vast complexity of the internet, and all in response to keystrokes from my mobile phone. As I lay in bed.

It's also worth noting the emergence of new kinds of value that are produced with scarcely any human input. There are many varieties of digital goods that bring value to a great many consumers without employing significant numbers of people at all. The most prominent are social media, a blanket term for many different services from Facebook to Twitter to LinkedIn. These are new kinds of goods, and economists are unclear as to how to account for them in their numbers (Bresnahan & Gordon, 1996). Two striking examples sum up this dimension of the problem, as we find fresh ways to create value that barely involve the labor factor at all. The old-style photography leader Kodak, which has essentially collapsed, once employed 145,000 people. Instagram, today's photography leader, had just thirteen employees when Facebook purchased it a couple of years ago for $1 billion. And to take the most stunning example, the communications sensation WhatsApp has been valued at $19 billion. It provides jobs for just fifty-five people. Yes, that's fifty-five.

One way to frame the emerging question is in terms of our creation of a new, intelligent "species" which is competing with us for work, and to which we are beginning to outsource our jobs – what Norbert Wiener (1950) termed the "precise equivalent of slave labor." As this process moves forward, more and more of our work will be able to be done by this new slave species, in its many forms. We do not need to posit huge strides in Artificial Intelligence for growing areas of what have been human jobs to be done better and cheaper by these machines. Nor do we have to assume that there will be no jobs in the future for people. But there is growing competition between people and machines for many of the jobs we have at the moment – and this is already being factored into discussions about raising the minimum wage, especially in fast food, which we discuss later in this chapter.

The key question to which we don't know the answer is whether new jobs will be created for humans in sufficient quantities to replace those taken by this new species. If so, what skills will be needed to do them? Given the likely pace of change (fast), the nature of the jobs being displaced (mostly, though by no means all, at the lower skills end), and the skills likely needed to compete with or work with the machines (considerable), the outlook is not encouraging. Everything depends on that set of assumptions. Will they bear the weight that is currently being placed on them, as leaders assume we don't face a significant problem?

The landscape of unemployment

There is nothing routine about structural unemployment. The assumption that there is skates too quickly over the actual effects of disruptive change in the past. And we do not need to go back to the first Industrial Revolution to find this out. We have examples all around us. Take the devastation wrought in mining communities in the UK by the effective end of coalmining in the past generation, after the infamous miners' strike of 1984 triggered dramatic reorganization and finally wind-down for the industry. One-quarter of the entire population of Wales had been dependent on coal-mining. In the US, the current municipal bankruptcy crisis in Detroit offers a grim example of the costs of change. As in cities like Gary, Indiana (named after one of the founders of the once-mighty US Steel, and now famed as a homicide capital of America), the collapse of traditional heavy industry has led to the sprawling "rust belt" – an amorphous, multi-state area in America's heartland where recovery has been very slow and is still a work in progress more than a generation after the collapse. So, it's important to take seriously the actual social impact and cost of what from a theoretical point of view may seem to be routine adjustments to technologies and accompanying global economic forces. Of course, it should also be said that these situations have been manageable, and that overall the economy – and employment – have done well as a result (Janoski et al., 2014).

To take an example from history, in a recent speech to the UK Trades Union Congress, Andy Haldane, the chief economist of the Bank of England, made the startling point that among the effects of the first Industrial Revolution was the steady doubling in size of the UK's unskilled labor force – from 20 percent to 40 percent between 1700 and 1850 (Haldane, 2015). Something similar happened a little later in the US. One effect of the new machines was thus to close down a great deal of skilled artisan labor and relocate it in factories and production lines. While new managerial and related high-level jobs did emerge, and there were wider social benefits through the growth of the economy, this process was also accompanied by a mass move not up but down the skills ladder. The point to note is a simple one: The actual costs of "structural unemployment" as we have already experienced it have been high.

This was noted back in the early nineteenth century by David Ricardo, one of the most influential economists in history. Reflecting on the impact of the Industrial Revolution, he expressed sympathy with the arguments of those who saw it as eroding the income and jobs of workers. But he did not favor pulling back from the technology, since that would just send capital and production abroad (he is famed as the originator of the theory of comparative advantage). He concludes that "the opinion entertained by the labouring class, that the employment of machinery is frequently detrimental to their interests, is not founded on prejudice and error, but is conformable to the correct principles of political economy" (Ricardo, 1821). In historical context, of course, we might locate this as an observation of the process of disruption leading to "structural unemployment," from which the economy (though perhaps not many of the disrupted individuals) finally benefited.

However, even accepting a basic analogy between past, present and future experiences of structural unemployment, the impact of the revolution in Machine Intelligence is unique because it is on the cusp of bringing about disruption in dozens of industries *at the same time* – and as part of a rolling process, the endpoint of which is not clear (or even if there will be one). This is illustrated in startling detail in the recent occupation-by-occupation assessments we discussed earlier. Previous disruptive experiences have been focused either on the invention of new devices (such as Hargreaves' classic Spinning Jenny) or the introduction of new general purpose technologies (think steam power), or the harnessing of the one to the other. What we have in prospect is a nascent technology that will tap each and every one of us on the shoulder and say, "How can I help?" And whatever it is we are doing, it will be able to help a little or a lot. Machine Intelligence may prove to be the ultimate example of what economists call a "general purpose technology" – one that has impacts throughout the economy. Steam power, and then electricity, brought about changes in industry after industry as their impact cascaded through the economies of the West. But that impact offers merely a foretaste of the potential of Machine Intelligence, which threatens to disrupt the human role in hundreds of different areas of employment.

The social devastation experienced in the steel mill communities of the American Midwest or the colliery towns of South Wales emerged in large part from the concentration of those communities on one single industry. When that industry goes down, down goes the entire economy of the community. Hence the description of Detroit as a company town that's now without a company. The localized nature of these disruptive experiences has made them much worse for the communities involved, but it has also removed them somewhat from the awareness of the wider community. At the same time, it has facilitated remedial interventions from governments anxious to retrain and redeploy the specialized labor forces whose raison d'être has evaporated.

In the case of the Machine Intelligence revolution, what we face is a disruption that mirrors previous disruptions but on a far wider scale. Its impact will not be localized geographically, nor as to industry. For two reasons: Partly because jobs passed directly from humans to machines will be lost in many different areas of the economy – the trucker laid off, and the checkout assistant, the personal banker, the legal researcher,

the college teacher, the telemarketer, the accountant...
But partly also for another reason: in many cases the impact of these new applications of Machine Intelligence will cascade dramatically and lead to the indirect loss of further employment that will not be geographically contained as it was when stores closed their shutters in the steel towns and coal-mining villages. This fact should help focus the attention of policymakers who have been accustomed to the geographical focus of previous instances of industrial decline, along with its heavy emphasis on unskilled and semi-skilled manual laborers. The prospect of the disruption to come suggests a much more demanding political situation since it will cut into the economic and social comforts of very diverse groups of workers, regardless of their political party affiliations.

Self-driving vehicles offer the clearest example, because their impact has attracted the most serious thinking, and as we have noted they are already being commercially deployed, if in limited contexts. As described in Chapter 1, in addition to the jobs of 3.5 million US truck drivers, the jobs of a further 5 million support staff are under threat; and those of cabbies – traditional and Uber – and bus drivers. With fewer vehicles purchased and many fewer built (back to the rust belt and its ills), the auto industry could be eviscerated. Add to this the possible casualties in the petroleum industry, from refineries to gas stations, in farming (ethanol), and in the auto insurance industry, the drop off in other modes of transportation, and the accompanying impact of increased safety (with the removal of a host of dangerous drivers from the road) on the demand for emergency physicians and nurses. It is inconsistent to make grand claims for the technological revolution under way and then to pull back into a business-as-usual approach to its impact on employment.

It goes without saying that new jobs will emerge as this remarkable shift works its way through our economic and social life. Already, we have hundreds of highly paid software and robotics engineers being hired by auto manufacturers and ride-sharing companies, as the traditional engineering needs for vehicles that despite their sophistication have continued to resemble Henry Ford's Model T are supplemented – indeed, transcended – by the requirements of autonomous, connected, intelligent vehicles that are more like smartphones on wheels. While ride-sharing companies Uber and Lyft plan to shift from human drivers using their own vehicles to fleets of autonomous vehicles, no doubt many fresh business models will emerge with new jobs to meet market demands at which we can currently only guess. But the two simple assumptions behind the scenario we are discussing – the replacement of human drivers (professional and amateur) by autonomous vehicles, and a corresponding shift from private ownership to fleet services with much higher levels of uptime/ usage – have profound implications for both the *kind of labor* needed and its *quantity*.

The ramifications through the economy of the application of Machine Intelligence to just one traditionally human function – driving – are immense, for the simple reason that our traditional technology has powerfully framed much of our behavior, and built up economic activity to enable it. Shift the core assumption – that humans are needed to drive cars – and everything is suddenly in flux.

That's why, rather than see the AI/robotics revolution as "yet another" industrial revolution like the first one, it may be more helpful to note the impact of the changes that are afoot in fresh terms. As Charles Murray has said, the case for "this time it's different" has a lot going for it. Clearly, at one level we are discussing the development of a new generation of algorithms and smart devices. But in terms of their effect on the industrial economy we may need to find a more effective metaphor. Should we be speaking about a new tech "species" able to take on our work? We don't need to claim that Machine Intelligence can do everything that human workers do – just much of it. And we certainly don't need to claim that humans will be incapable of coming up with completely novel things to do with their time once these tasks are assigned to machines, although how those activities will relate to the traditional economy is not at all clear.

It's important to remember the pace of change, and for policymakers to anticipate potential scenarios. Of course we can expect new jobs to emerge – they do so all the time at the meeting-point of human creativity, social need, and economic opportunity. But these things take time, they may require the development of new skills, and in parallel we may expect continuing, competing, fresh applications of Machine Intelligence. What kind of educational needs – for children and students, but also for the rest of us – will be needed? These are urgent questions, in respect of both new kinds of jobs and the potential for much greater "leisure." We have no reason to expect a pause – a plateau – to enable skills to catch up, new occupations to develop, and economic opportunities to emerge in the quantities that will be needed to absorb large numbers of workers whose jobs have gone to machines. If we are "racing" against the machines, they keep moving faster – a point that Keynes (1931) himself noted as the core problem: "Unemployment due to our discovery of means of economising the use of labour outrunning the pace at which we can find new uses for labour."

An increasingly comprehensive range of applications of Machine Intelligence is already evident, so much so that humankind's most profound invention will evidently be another de facto "species." Not a humanoid with "Artificial General Intelligence"; indeed, not a humanoid at all. The point is that we are amassing a collection of applications that cover a growing slice of what humans can do, task by task, and that is therefore best understood in these sweeping terms: as the creation of a machine alternative to humans – capital substituting for the labor factor in production. The machine "slave species."

I remember this point being forcefully made a decade back by computer scientist and writer Marshall Brain, who became famous for founding one of the earliest commercially successful websites, Howstuffworks. He was speaking at a conference of the Kurzweil-associated Singularity Institute, and was underlining the problems we would face down the line when these Artificial Intelligence applications spread across the labor force. He put up on the screen the US Department of Labor job classification form, and went down the list of categories that would be hit; he came up with a total of around 50 percent, which pretty much anticipated the more detailed study by the Oxford economists Frey and Osborne. The audience consisted largely of technology enthusiasts, and his cautionary words were not well received. Perhaps it was no surprise that he was heckled. One particularly barbed remark stays with me: "Are you aware, Professor Brain, of an academic discipline called history?" In other words, because in the past disruptive innovation has led to more and better jobs, the future will be the same, so to raise such problems is foolish. Which takes us back to the fulcrum question: Is this time different?

New sources of employment?

The "standard" answer is perhaps best encapsulated by Artificial Intelligence guru Ray Kurzweil. His position lies at the extreme optimism end of the scale on two different sets of issues – both the pace of change, and whether this will be good news for us. As the website of the Kurzweil-related Singularity University states, in response to "a growing concern about what people are going to do when all the regular jobs are done by robots, Kurzweil put it like this: 'We are destroying jobs at the bottom [of the] scale ladder. We add new jobs at the top of the scale ladder. The scale ladder moves up. In order to keep up with that rising scale ladder, we need to make people more skilled.'" Citing similar concerns about the introduction of machines into the textile industry at the dawn of the Industrial Revolution, Kurzweil said, "You could point at almost every job and it seemed only a matter of time before those jobs were automated and eliminated. Those jobs were automated and went away. Yet somehow, employment went up... New industries emerged making and servicing those machines" (Hill, 2015). The article concludes with the haunting question, "What are humans uniquely suited to do?"

Lawyer Garry Mathiason, whose firm specializes in robotics employment issues, says much the same thing. He notes that this is basically an economic issue – and opines that pushing up the cost of labor through minimum-wage increases is plainly going to make using robots more attractive (for example in New York fast-food outlets). Bill Gates makes the same general argument: Government should encourage employers to use people not machines, and pressing for a higher minimum wage will have the opposite effect. Mathiason looks ahead to the impact of the growing movement for higher minimum wages, which "will have the effect of making it economically more attractive to bring in robots to absorb some of that work. I think

The Lock on Twitter

that will accelerate the speed with which this will take place. So over the next five years, we're going to all of a sudden see robots be very much a part of our lives just in terms of our normal commercial activity" (Giang, 2015).

But he's not worried: "There will be a displacement and there will be a repositioning of people into jobs that we don't even have today that we will have in the future. If you look back in history, you'll see that this disruption has been going on for some time. Not as fast as what we're currently experiencing, but nonetheless there." And he gives the classic example, that of US agriculture. In 1870 it still employed 70 to 80 percent of the population. Today that figure is less than 1 percent. What proponents reiterate is that in earlier situations, from the first Industrial Revolution on, while new machines have displaced workers they have also led to the creation of new jobs – whether working on the machines or thrown off by them into the new economy they bring into being. Easy examples in the context of the digital revolution of the past generation would naturally include work in the trenches for the technology companies, but also such roles as Search Engine Optimization, the kind of job no one could have foreseen would exist; selling goods through eBay; driving for Uber and letting through AirBnB; and blogging (from which, surprisingly, some people make good money!).

Mathiason certainly foresees huge changes, and he's looking at them on behalf of his law firm. "Where will wage and hour requirements stand for robots...? Will they receive minimum wage and overtime pay requirements?" (Giang, 2015).

The argument that new job opportunities may be expected to emerge is well put by *Forbes* magazine writer Steve Denning. Critiquing Martin Ford's recent book *Rise of the Robots* he argues that "when machines replace human capabilities, as they have for thousands of years," it isn't that "nothing else changes." "In reality, as Philip Auerswald...author of *The Coming Prosperity* (2012) points out, when machines replace one kind of human capability, as they did in the transitions from hunter/gatherer, from serf, from freehold farmer, from clerical worker, from knowledge worker on to what comes next, new human experiences and capabilities emerge" (Denning, 2015).

He continues: "Often these new experiences and capabilities were unimaginable in the prior era. The new experiences and capabilities were mostly higher value, and offered more interesting work, than the experiences and capabilities that had been replaced by machines. This was true of agriculture, industrialization, mass production, and so on. Why should it be different now?" That final sentence of course takes us back to the heart of this debate. Is it different or not? Or, better, can we be *sure* it isn't different? To return to the risk-framing of the question, a prudent policymaker might ask: Given the stakes involved in potential huge disruption to twenty-first-century labor markets, we really need to be very sure that it won't be "different this time." And what's more, if it's "the same" – if what we face is essentially a replication of previous disruptions – the policymaker will ask what "the same" means for the AI/robotics revolution. Because "the same" may prove to be very "different," as the impacts of Machine Intelligence ripple through probably every single element in the labor market.

One thing that seems plain is that we need not understate the smartness of machines when it comes to taking over functions that are currently done by humans. Denning himself notes that the "new experiences and capabilities" that emerged from earlier innovation revolutions "were unimaginable" beforehand. This of course cuts both ways. It's true that we cannot imagine some of the new jobs that may be on their way, any more than we can imagine a social order in which many more people don't work for a living. On both sides of that argument it pays to be cautious. As recently as 2004, Frank Levy and Richard Murnane were convinced that driving a car was so complicated that it would be unrealistic to expect robots to do it in the foreseeable future (they single out the challenges of a left-hand turn against traffic in right-hand-driving countries). As we know, it wasn't long before Google showed them to be mistaken.

It's worth underlining the point, illustrated by the wide range of efforts we are summarizing and the many more we are not, that we do not need the advent of "Artificial General Intelligence" – a robot with a brain as smart as ours and therefore all-round skills – for many or perhaps most of those tasks that our human smarts enable us to perform to be taken over by individual machines. Submarines, as Jerry Kaplan put it, don't swim, and ATMs dispense cash (and in some cases receive our deposits) without smiling or asking how we are. To put it bluntly, it would be unwise to bet on any particular human function being "secure" – safe for our species to perform, safe from the rivalry of machines. So, in reviewing Geoff Colvin's lively and perceptive book *Humans are Underrated*, writer Frank Dillon (2015) quotes: "The pattern is clear. Over and over, smart people note the overwhelming complexity of certain tasks that humans do, and conclude that computers won't be able to master them, yet it's only a matter of time before they do."

Wishful thinking?

The prospect of a major disruption involving large-scale structural unemployment might be unlikely – but only if one (or both) of two assumptions can be made.

First, if the future is one in which the impact of Machine Intelligence on existing employment is significantly less than many have predicted. This could be for various reasons. Perhaps the predictions are simply wrong and the Oxford and Pew studies should simply be discounted. Perhaps the technology will take much longer to be applied. Or perhaps we humans will just prove more resistant to the handover than predicted. Perhaps we decide to resist driverless cars, or demand personal service – in supermarkets, banks, fast food restaurants, across the economy – and are prepared to pay extra for it. Perhaps we will be suspicious of handing over care of the elderly to machines. Perhaps we will place a premium on both services and manufactured goods that are "artisanal" right across the board, and keep alive the market for human labor even in the face of cheaper machine alternatives.

Or, to put it another way, perhaps we will see something akin to the remarkable European repudiation of genetically modified foods in the late 1990s. In a startling rebuff to the biotech giant Monsanto, European consumers rejected these products and essentially demonized them, such that twenty years later questions such as GMO labeling and the legality of growing GMO crops remain highly contentious issues in Europe and beyond.

We could perhaps frame this as a "Soft Luddite" response (and no doubt some techno-boosters would label it like that!), though it could have many contributory causes. We may wish our old people to have real people to comfort and tend for them. We may actually like real people serving us in stores and banks and restaurants. We may be genuinely happy to pay extra if it keeps someone in a job. It's perfectly possible to buck the economics of a new technology, or appear to, if we value more heavily what economists have discounted. Some consumers will pay a lot more for organic foods. A consumer preference could extend to the personal and the artisanal in contrast to machine options. The impact of such a movement would be to place significant constraints on this particular technological revolution.

A second justification for discounting our concerns, perhaps more realistic than the first (and more in

81

tune with the view of those techno-boosters who have addressed the question), would in fact require three separate components in the scenario for it to work.

First, the rapid arrival of new jobs. No one doubts that there will be new careers emerging, and there may be many. The issue is whether they arrive soon, or more slowly and over time. The latter possibility does not require any particularly dystopic thinking on our part. It's simply that the move from, say, coal-mining or steel manufacture to Search Engine Optimization or eBay selling, or whatever is regarded as the quintessential job of the "new economy," took time. The pace of change on the new-job front needs to pretty much match the pace of change of the disruption to avoid a serious "structural unemployment" quotient. If the process reflects how long it took to shift out of the traditional heavy industries, the outlook is not good.

Second, it involves the assumption that the new jobs will come in diverse and large quantities. Plainly, if the target we need to hit – as per the Oxford study – is 47 percent, mass employment opportunities will be necessary. If it is initially only 9 percent, as per OECD, the scale of the challenge is much less. If 3.5 million truck drivers need to be redeployed, there will have to be 3.5 million vacancies for jobs XYZ. Or the equivalent, across the economy. From our experience of the creation of new kinds of jobs in the current economy, this requires quite an optimistic exercise of imagination. While all developed economies are facing challenges, the stagnant or declining state of middle-class incomes in the United States, and the drop of 10 percent in overall incomes in the UK since 2008, create uncertain contexts for these fresh and potentially dramatic disruptions.

Third, the education and training programs required to convert workers in the current economy into candidates for these diverse new opportunities need to be effective and timely. This may be the greatest stretch of all. Even if we knew what was needed, effecting fundamental change in public education (in the US, for example, though the situation is similar in other major developed nations) is a controversial, and glacially slow, process. Our leaders need to begin to think on new, fast-moving timetables. The education/skills training implications of engaging seriously with the disruption scenario we are discussing stand out as immediate challenges for policymakers thinking ahead. The industrial economies have never experienced anything like it. I'm haunted by the comment from Wiener that it could make the Depression of the 1930s "look like a pleasant joke."

We don't know what comes next, but what we do know is that some of the best brains on the planet have spent decades wrestling with how to get machines to do stuff that in the past only people were able to do. Their goal was once largely academic, but as progress has been made it has become increasingly economic. To that extent developments in the Machine Intelligence field mirror the story of the internet – originally an academic project, publicly funded, but growing into an enabler of economic activity and a source of great corporate and personal wealth for those who succeed in harnessing its commercial power. Many of the world's top experts in Artificial Intelligence now work for companies like Google. Toyota, the world's largest automotive company, is establishing a billion-dollar robotics research institute split between the campuses of MIT and Stanford. Economic motivation is now in the driving seat.

The search is on, in industry after industry, for cheaper and more efficient methods of delivering value – methods that do not involve people. Intelligent machines can operate 24/7. They don't join labor unions. In fact, they avoid altogether the need for an HR department. And while the impact of the new automation revolution varies greatly across industries (for example, the Amazon Mechanical Turk, along with various forms of part-time, female-dominated and non-unionized labor, demonstrates the continued appeal of cheap, low-skilled labor), its pervasiveness needs to be noted; unskilled areas such as agriculture where labor rates are low have already become beneficiaries of some of the most imaginative Machine Intelligence applications, all the way from highly computerized John Deere tractors to complex robot-like harvesting machines for crops that were until recently regarded as incapable of being mechanically gathered.

How we handle the question of the distribution of income and wealth will determine, from an economic point of view, how people will fare if the labor factor disappears from the creation of value.

Towards a disruption consensus

However things are finally resolved, what lies on our horizon is the prospect of disruption. It's around this that a new consensus needs to be built and a more realistic "conventional wisdom" needs to emerge. From one point of view, it's just a more serious restatement of what the techno-boosters have already been saying, if somewhat thoughtlessly. They pay lip-service to the big changes that lie ahead, but refuse to admit the scale of the revolution that's in the making. They acknowledge – at least implicitly – that we face the onset of a new wave of structural unemployment; that's what all major technological upheavals bring about. But how extensive will this bout of unemployment be? How long will it last? Will it be comparable to the decline of the heavy industries, or something the like of which the economies of the modern world have never seen?

The responsible thing is to plan for all outcomes that are seriously possible. That's how major corporations work, and it's also how governments work, when they are well led. On another scale, it's also how we work as individuals, if often intuitively, when we are making our own plans for the future. If we aren't quite sure how things will be, we prepare for all eventualities – and make sure that we are not blindsided by a surprise outcome. None of us expects our house to burn down. But we buy insurance.

The evidence is clear that the dislocation brought on by the diffusion of Machine Intelligence through our labor markets will be considerable. That is all we need to argue. Can that give us a basis on which to build consensus?

FOUR

Building Consensus and Getting Prepared

The need is for a policy consensus that is accepting of a range of diverse predictions – from the dawn of a workless world to the prospect of plentiful new jobs to sustain the "full-employment" model. It needs to focus on preparing for the likely turbulence ahead, but also accept the possibility that more jobs may be lost than are created, with a dramatic cumulative impact in the long term. Whether the anticipated job losses are closer to 9 percent or 47 percent, it's prudent for policymakers to be prepared. If the pattern of previous industrial disruption is repeated, serious structural unemployment and the emergence of "rust belts" await us. But how to prepare?

To those who sweat for their daily bread leisure is a longed-for sweet – until they get it.

John Maynard Keynes (1931)

We began by suggesting that, since we don't know what's going to happen, we need to ask ourselves two basic questions. First, on the conventional assumption that new jobs will emerge to take the place of those that go to machines, what kind of labor market turbulence can we expect during the transitions that will be involved? Second, is the idea of big job losses that aren't replaced with new jobs just ridiculous, or a serious possibility?

Note that neither question assumes what David Autor (2015) has called the "dystopian" view of massive job losses. The first question assumes a major impact on labor markets as advanced applications of AI are rolled out. This is a view commonly held, and (as Charles Murray notes) even the OECD's "low" figure of 9 percent suggests big changes ahead, hence the "rust belt" scenario, to be expected in the various sectors of the economy where the applications we have called "non-human resources" begin to be applied.

The second question addresses the "dystopic" labor market scenario in the context of risk. What is the chance of a major meltdown that, at least in the short to medium term, is incapable of being halted? This Keynesian hypothesis, which now has a growing range of contemporary backers, is alarming. What it suggests is that the next disruption coming from innovative technologies will involve progressively reshaping the factors of production to exclude human labor – from some factors, then many, then most. But the question is focused not on our take on what will happen (on which there are many views), but on the risk that this could happen.

Since the future is always in varying degrees uncertain, corporate strategists engage in "strategic planning" which seeks to manage that uncertainty by expressing it in terms of risk. How likely is X to happen, and if it were to happen what would the impact be? The first task is to quantify the risk (how likely?) and reckon its potential impact (how big?). The second is to strategize how best to make X less likely, and then, finally, how to limit its impact if, despite our best efforts, it happens. This is the case with any scenario that would have a significant impact, and that is not considered so implausible that it can be disregarded.

Of course, as in corporate contexts, risk assessment and management are complicated by the fact that in order to cope with the possibility of X we must seek to avoid causing damage elsewhere. We don't want to end up with a cure worse than the disease. That is part of what motivates those who are critical of any effort to plan for the possibility of X. They argue that raising such questions about the potentially negative impacts of innovation will dull its edge and mitigate against its success. This does seem a naive way of handling the process of exponentially driven technological change.

As MIT economist David Autor (2015) notes, arguing against what he refers to as "Recent dismal prophesies of human-machine substitution," they typically fail to address the implication of their argument that if things turn out that way "then our chief economic problem will be one of distribution, not of scarcity." It's worth noting because it bridges directly from the arena of economics into that of policy. If there is merit to any other than the most rosy scenarios, the question of distribution in an economy in which capital is taking a greater role and labor a lesser one is unavoidable.

Prudent responses to the two basic questions raised above should lead us into consensus-based contingency planning, whether as individuals, organizations or governments. Everyone agrees that we face disruption ahead, and that it's prudent to prepare. While timelines are the most uncertain of things to predict, the disruption may be coming soon. The Oxford study is appropriately vague – but speaks of "a decade or two."

So how to prepare?

Preparing the public

When I was asked recently by a government adviser what we should be doing, I first responded that our leaders need to frame the questions for us. It's time for speeches. We have yet to hear a single major political leader anywhere on the planet devote a speech to the prospect of a coming labor crisis. The nearest we get is the occasional comment on the need for workers to get their fair share of the benefits from technology in the "new economy." (President Obama's final State of the Union address included a couple of remarks along these lines.) But this is to misunderstand the problem. While it may be true that technology is enabling corporations to squeeze more out of their workers, this is a transitional situation. The issue is not going to be whether workers share in the benefits of the enhanced productivity enabled by machines. The issue will be whether they are still workers.

So the first need is for leaders to frame these questions, and to do so in a manner that matches their seriousness and urgency. But the problem leaders face is that the questions don't fit the political agenda – anyone's political agenda, at least within the mainstream. The need – and this is why leaders are so crucial when transition is in the wind – is for the issues to be framed, and framed aright. What's coming is a situation in which we may expect fundamental shifts in the industrial economy we have inherited from the twentieth century, an economy whose structure goes back into

So much else happening with Trump

the nineteenth and eighteenth centuries. It's all change, because "full employment" may be over. Just as we are getting used to the idea that people change jobs more often, and that the notion of a job-for-life perhaps within one company is a thing of the past, the idea of there being a full-time job for anyone who wants one may be going the same way. We could be moving towards a global rust belt, in which the skills possessed by workers simply don't match the (diminishing) number of opportunities for skills to be reimbursed as jobs. Because, above all, we find ourselves competing against members of another species, a machine species, cheaper to employ, and constantly evolving into smarter forms.

So we need leaders to frame the discussion, to prepare their various constituencies, to alert them but also to educate them. Because while this emerging situation needs to be seen as a coming crisis, it also presents a possibility of unparalleled opportunity.

The problem is partly one of locating this vast challenge on a recognizable agenda. One reason why labor leaders – whom one might have thought its obvious standard-bearers – have been reluctant to be drawn into the discussion lies in their hesitancy to oppose innovation and resist technology. Not just because they fear being labelled Luddites, but because the lesson of the past is that however painful transitions have been, the end result has indeed been greater prosperity and increased employment opportunities. And it may in fact be helpful that we are not entering this discussion with loud voices framing it in fearful terms and pressing for us to resist the new technologies.

Yet, one way or another, we face the need for an informed public understanding of the issues we are discussing, as the background for a maturing of both personal and political decision-making about the future.

Preparing the government

My second suggestion to the government adviser was to do with policy. What we need is a comprehensive review of every policy area to prepare for a situation in which the labor market faces turbulent disruption – a situation in which, in traditional terms, we face rising unemployment for structural reasons. Whether we are looking at the erosion of half the jobs in the economy over a couple of decades, or many fewer, what are the displaced workers going to do? The examples of South Wales, Detroit and Gary show that while new jobs do indeed arise, the process is messy and protracted, and may well not generate enough opportunities for those individuals whose skills are no longer relevant.

So, what would a policy regime look like – in the US, the UK, or anywhere else within the OECD group – if unemployment were not around 5 percent (close to "full employment"), or 10 percent (a near-crisis level that is a threat to governments but is close to the current EU average, weighted by the much higher levels in southern Europe), but, let's say, 30 percent? What policies would democracies need to have in place to face an extended period with unemployment at that level? Such levels are characteristic of the localized cases of severe structural unemployment we have discussed, and they have persisted for many years before slowly being reduced.

For the sake of argument, let's take certain policy objectives for granted. First, that it is a social good for people to be occupied, whether part-time or full-time, whether in "employment" or regular voluntary activity. Second, that there needs to be reasonable provision for the sustenance of all citizens. This may be seen either as a social good, or as a necessary evil; and very different approaches are taken in different nations, with a sharp divide on the two sides of the Atlantic. But – and this becomes especially important as unemployment rates rise – there is also a simple economic issue at stake, which is the importance of demand.

New significance will attach to the notion of "full employment," and the political importance of the "unemployed" percentage, as the significance of the human factor in producing value declines and the roles of capital and technology grow. The legacy of these ideas from the industrial society of the twentieth century will ebb, and with it many attendant ideas such as the full-time/part-time distinction. That, like much else on this agenda, will be of greater significance in the United States than in many European societies since in the US health and other benefits tend more to follow full-time jobs. But it's a key distinction in all industrialized economies, which typically set an assumed number of hours worked as the basis for various policies and benefits (such as retirement income and parental leave). This set of assumptions requires re-examination in a context in which there is less work to go round, and there may be a policy imperative to encourage the sharing of work (for social reasons) rather than its being contained within full-time "packages" wherever possible (which is the current default).

Plainly, the same is true of the employment/self-employment distinction. The issues here have been discussed, especially in the United States, in the context of the emergence of what has been called the "gig economy." While the shift so far may have been exaggerated, it has long been problematic that the self-employed are treated as a subordinate tier of the labor force, though the emergence of Uber and its use of contractors who are de facto employees has given it some fresh bite in policy and in the US courts.

Something similar applies to the distinction between voluntary work and unemployment. Voluntary efforts – while seen as socially valuable by governments, and often funded, directly or indirectly, as a means of delivering social services – are largely off-the-grid from the policy perspective. And again: the distinctions between being "employed" and being retired, and between employment and education. Each of these derives from the traditional "full-employment" assumptions of twentieth-century industrial society.

Meanwhile, as jobs are leached from the human economy, governments will endeavor to do all they can to persuade employers to keep people in place. Commentators such as Bill Gates have already noted that the push to raise the minimum wage will accelerate the use of machines in fast-food and other low-wage jobs, and the question of simple competition between a human job and the slave machine that can be "hired" more cheaply may become a major political question. Governments may decide to cut back on the costs to employers of retaining human employees – through taxes and benefit obligations – in order to maintain jobs. They may also begin to tax the hiring of intelligent machines to replace human workers – both to level the playing field and simply to replace lost revenues as human jobs are displaced. Such moves may be seen by techno-boosters as Luddite efforts to assault the new technology. But governments are constantly seeking to manage both levels of employment and the tax take, and there is no reason to expect them to change their behavior once these changes begin to bite. One reason why they need to start thinking much further ahead on this set of issues is precisely to enable them to adapt appropriately to the new circumstances that are evolving. Recent global challenges in relation to the taxation of "new economy" companies such as Apple, Amazon, and Google illustrate another aspect of the same problem (Kennedy, 2016).

These considerations shed fresh light on some of the current controversies involving disruptive technology applications, such as the Uber cabs. Efforts in some cities (such as Paris) to bolster the ability of traditional cabbies to compete by requiring a delay in Uber pick-up times may seem amusing attempts to stave off progress. But they prefigure what are likely to be widespread government efforts to manage the transition into the machine replacement of workers.

Universal income?

A core policy proposal that has been under discussion for many years and is beginning to take on fresh life is that of a Universal Basic Income (UBI). As the need for human labor to twin with capital in order to produce value diminishes, the link between income from employment and personal sustenance is coming under threat.

There are differences of emphasis within the developed economies in respect to both the level of the "social wage" (the amount of national income the government provides for services) and the level of support granted to the poor, under-employed, and unemployed (in some countries it differs little from what they would earn if they worked full-time in reasonable jobs, in others it is more meager). But there is no disagreement that if levels of unemployment rise, not only will social (and indeed political) problems be on the agenda, but there will also be a drop in demand, and therefore further pressure on the economy. In response to this set of concerns, an idea originally developed as a means of radically simplifying welfare benefits has been revived – interestingly, by economists and politicians of widely differing perspectives. In that sense, while the idea of a Basic Income may look like a "socialist" solution to the challenge we are facing, since it is fundamentally about the redistribution of national income, it deserves a political branding all of its own. One major advocate was the conservative economist Milton Friedman (Orfalea, 2015):

> We should replace the ragbag of specific welfare programs with a single comprehensive program of income supplements in cash – a negative income tax. It would provide an assured minimum to all persons in need, regardless of the reasons for their need...A negative income tax provides comprehensive reform which would do more efficiently and humanely what our present welfare system does so inefficiently and inhumanely.

The push for this idea to be given fresh consideration comes in part from a view of what is likely to happen otherwise. If we carry our present policy regimes forward into a situation of steadily rising unemployment as we enter the first stages of the anticipated disruption, then we will face increasing inequality and major social disruption as the rust belt spreads. Basic Income advocates have begun to press their case with new vigor, most recently Charles Murray (2016) in the pages of the *Wall Street Journal*. As conservative versions tend to be, Murray's proposal is especially radical in that he wants it to take the place of all social benefits (including social security, the US retirement program). What's interesting is that he links it to the labor market question – it's "the best way to cope with a radically changing US jobs market." After reviewing the Oxford and OECD reports, he concludes: "Even the optimistic scenario portends a serious problem. Whatever the case, it will need to be possible, within a few decades, for a life well lived in the US not to involve a job as traditionally defined. A UBI will be an essential part of the transition to that unprecedented world."

Arguments pro and contra UBI will continue, but if the labor market is soon to suffer the kind of turbulence we have discussed, it's hard to see how politically acceptable solutions will be able to avoid more solid provision than is currently available for the unemployed and under-employed (and those involved in retraining). UBI is not the only way of bringing that about. But it becomes increasingly attractive as scenarios become more dramatic.

Meanwhile, various European nations are beginning to experiment with versions of UBI. The agenda is in general a broader one of social welfare reform rather than an immediate effort to address the themes discussed in this book. But continuing interest from

100

who are?

various parts of the political spectrum is keeping the UBI concept alive, not least at a time of burgeoning welfare budgets. Switzerland recently voted one version down, Denmark and Finland are among countries that have run trials. Clearly, with their higher "social wage" and in general more generous policies in relation to the long-term unemployed, European nations may see themselves at a competitive advantage vis-à-vis both the United States and developed Asian economies as the anticipated structural disruption arrives.

Preparing the workforce

The third set of implications for government plainly lies in the realm of education and training. If we are hopeful of the emergence of new jobs that will survive rivalry from the new, evolving species of intelligent machines, great weight attaches to our preparing the best possible strategies for educating and re-educating the workforce.

The orthodoxy at the moment is to focus on STEM (science, technology, engineering, math) education, as a means of better preparing the rising generation for the new skills marketplace in an economy in which technology is of growing importance. While this has value, we have already noted that it is not without a core problem: As every projection of potential machine impact on the job market makes clear, many STEM-related jobs are likely to go to machines sooner rather than later; and the (at the moment!) "uniquely" human characteristics are little enhanced by these technical skills. This is not to suggest they are irrelevant to preparing students at present, nor that we do not all need to be better schooled in skills that will enable us to understand and engage with the machines. But, plainly, STEM skills are not the elixir. What we do know is that those distinctively human capacities that will prove harder to replicate lie in such aspects as intuition, relational abilities, and creativity.

There is also the crucial issue of adaptability. It has become a cliché to say that we need to be prepared for career shifts in the future. In fact, looking ahead, our children will need to be almost infinitely adaptable to manage lives that will likely be increasingly long and may include variants on full-time and part-time "work," both employed and self-employed, voluntary work, periods of leisure time, and more. What will prepare them well? Plainly such skills as a capacity for self-invention and re-invention, the ability to fill their time fruitfully on their own and without either supervision or the pressure of economic need, a life of the mind, a capacity for relationships of many kinds, perhaps above all a capacity to manage, cope with, even enjoy change. This is far from a naive STEM prescription. Insofar as it correlates with our current curriculum it lies squarely in the domain of the liberal arts and humanities. But more fundamentally it will call on inter-disciplinary capacities, on emotional intelligence, on deep relational skills. The acronym STEAM, adding "Arts" to the mix, offers a much more realistic notion of the skills needed as we move forward: Engaging with technology will be crucial, perhaps mediating between technology and its users, but human creativity and empathy will be central to careers in a work environment dominated by Machine Intelligence. Clearly, the educational implications of this new situation will require much forward thinking.

Preparing ourselves

How to prepare ourselves and our families for the coming disruption is no easy question. As students mull career choices, they may wish to consider how likely their preferred option is to survive the machine onslaught, and the respective merits of the various possible applications of their skills and interests. There are also strategic decisions to make. Double down on STEM skills with the aim of being among those who build and shape and maintain the machines? Focus on issues at the interface of machines and humans, and the competencies that will arise at the meeting-point of what in Spielberg's memorable movie *A.I.: Artificial Intelligence* are called "mecha" and "orga"? Or go for careers that have a high survivability quotient? Top of the Oxford list for survival is recreational therapist. At the bottom – #702 – is telemarketer. It's doubtful anyone will mourn the passing of telemarketers, except those unfortunate people whose career it is now, though dealing with machine marketing calls may only make the experience worse – but no doubt we shall have bots to receive the calls for us.

Quite apart from the near-absence of this discussion in the policy arena, it's amazing it is not a constant talking-point in our high schools and colleges and grad schools. And not as an add-on, an elective of "jobs and tech," or an occasional career advisement session. Every one of these institutions operates on the tacit understanding that there are no big changes on the way, that while careers are becoming more fluid and workers more flexible, their broad outlines will remain the same. In the process they are shamefully misleading young people. The choices they make today will prove fateful for how well- or ill-prepared they are when the labor market gets increasingly turbulent. There's a general acknowledgement that flexibility will be more important in the future, but that's a lame understatement. In our great educational policy discussions – focused especially in both the US and the UK on national legislation and big government interventions in the direction of K-12 schooling – there's hardly a word about what may emerge as the biggest issue of the future.

There is also the question of leisure. As tech guru Stowe Boyd puts it: "The central question of 2025 will be: What are people for in a world that does not need their labor, and where only a minority are needed to guide the 'bot-based economy?" (Smith & Anderson, 2014). The question of education for leisure may sound strange. But it redirects us from the contemporary discussion that is almost entirely focused on job preparation to the root purpose of education for human flourishing. Now is the time to start asking fundamental questions about its purpose, and how we should best be preparing both young people and the rest of us for the strange world ahead. As someone once asked me after I had spoken on this question, what should be the content of our education if on graduating from college at the age of twenty-two the student then retires? The question could not have been better put. Keynes' "new leisured class" has so far proved a pipedream. But the AI revolution could rapidly bring the questions it raises into sharp focus.

We have explored some of the decisions that governments will need to make if the decline in traditional employment is to result in something other than a growing class of unemployed poor. But whether or not the disruption results in greater inequality, it is likely to mean there are fewer working hours to go round. In other words, whether we end up poor or rich, we are going to have more time on our hands. It's worth noting that this "time" has been a major driver of social ills in the communities of both the twentieth and, so far, the twenty-first century. The old saying that "the devil makes work for idle hands to do" has been proved beyond a doubt. Even with subsistence income provided, as is the case in most developed countries, the experience of unemployment – for the individual and the family – has often led to social breakdown and crime (Ajimotokin et al., 2015).

It's also notable that among well-educated business and professional people the experience of unemployment for more than a short transitional period can be devastating. Since their experience offers one clue to the problematics of our future, it could usefully be studied with that in mind.

Aside from demonstrating how difficult it is to make a big difference to what goes on in our schools, the recent efforts of governments to improve the delivery of public education have been essentially directed to better preparing students for the labor force of the twentieth century. What of the twenty-first? We need to re-think our curriculum and, while we are not here offering a prescription, this looks like very good news for the humanities and liberal arts, which offer as much instruction in living as they do preparation for working, but within a context far more aware of future developments, our engagement with technology, and especially the interface with the intelligent machines who will increasingly populate the work and leisure lives of members of *Homo sapiens*. For a start, we might use literature and film to set the study of "science fiction" at the heart of the educational enterprise, permeating the curriculum.

The twenty-first-century pace of change demands a completely different mindset from the aging model that well served the old industrial economy. And as we gear up for a potential tsunami, what we need now is a massive effort in education for the rest of us: the same kind of education. A parallel retraining of the workforce, so, as the waves begin lapping, we shall be better prepared to handle what is coming. The advent of MOOCs may offer one simple path into such a project, even though it will not create many new jobs for teachers in the process.

The task we face is extraordinarily challenging, but the promise – if we can harvest the fruit of AI for the good of the human community – is unimaginably great. Alongside the brilliance of the mathematicians, engineers and entrepreneurs who are enabling these extraordinary developments, we need the very best efforts of policymakers, public intellectuals and political leaders to help us navigate our way into a very different future that is bright with promise.

Looking back and looking ahead

Once upon a time, I was a student at Cambridge University. I was a debater, and in my first term was invited to speak at the "freshman's debate." It was a memorable experience for more than one reason. The President of the Union that term was an outspoken Greek woman of conservative views and, apparently, significant wealth. Her first book appeared soon after graduation – an attack on contemporary feminism. Her name was Arianna Stassinopoulos. We know her better now as Arianna Huffington. Like many of us, she has shifted her convictions a little since those far-off days.

I've no memory of what we said that evening, in our teenage tuxes and ball gowns, after the sherry and before the more serious drinks, but the topic has haunted me ever since. For some reason I saved the poster promoting the event, and more than forty years later it stares down at me from the wall as I write.

"This House Believes That the Price of Progress is Too High."

But I spoke against the motion.

Introduction: Time to Stop Being Naive

1 Historians generally agree that there never was an actual Ned Ludd. The original reference seems to be to the English town of Ludlow in the year 1811, where a rash of "machine-breakers" destroyed stocking-frame machines, and employers began getting threatening letters signed by "Ned Ludd." Machine-breaking itself was not new (Thomis, 1970).

1 Non-Human Resources

1 Details on Rosalind Picard's remarkable work on affective computing can be found at http://web.media.mit.edu/~picard/publications.php.

2 "The Stupid Luddite People"

1 The Luddite slur on Elon Musk comes from the Information Technology and Innovation Foundation. Maybe it's unfair, but I detect a concern to ward off such an accusation in the over-the-top subtitle of

Notes

Brynjolfsson and McAfee's *The Second Machine Age* (2014), with its focus on the technological unemployment agenda: *Work, Progress, and Prosperity in a Time of Brilliant Technologies*. The title of their earlier volume *Race Against the Machine* captures the issue much better. For more on the warnings about AI from Musk, Hawking, and Gates, see Sainato (2015).

Ajimotokin, Sandra, Alexandra Haskins, and Zach Wade (2015) "The Effects of Unemployment on Crime Rates in the U.S.," smartech. gatech.edu, April 14.

Arntz, Melanie, Terry Gregory, and Ulrich Zierahn (2016) "The Risk of Automation for Jobs in OECD Countries: A Comparative Analysis," *OECD Social, Employment and Migration Working Papers*, oecd-ilibrary.org, May 14.

Auerswald, Philip (2012) *The Coming Prosperity: How Entrepreneurs are Transforming the Global Economy*, Oxford: Oxford University Press.

Autor, David H. (2015) "Why Are There Still So Many Jobs? The History and Future of Workplace Automation," *Journal of Economic Perspectives*, Vol. 29, No. 3.

Bessen, James (2015) *Learning by Doing: The Real Connection Between Innovation, Wages, and Wealth*, New Haven: Yale University Press.

Birkett, Kathy (2014) "Robot Caregivers," seniorcarecorner.com.

Bresnahan, Timothy F. and Robert J. Gordon (1996) "Introduction," in *The Economics of New Goods*, Chicago: University of Chicago Press.

Brynjolfsson, Erik and Andrew McAfee (2011) *Race Against the Machine: How the Digital Revolution is Accelerating Innovation, Driving Productivity, and Irreversibly Transforming Employment and the Economy*, Lexington: Digital Frontier Press.

Brynjolfsson, Erik and Andrew McAfee (2014) *The Second Machine Age: Work, Progress, and Prosperity in a Time of Brilliant Technologies*, New York: Norton.

112

Bibliography

CB Insights (2016) "33 Corporations Working on Autonomous Vehicles," cbinsights.com/blog, August 11.

Colvin, Geoff (2015) *Humans Are Underrated: What High Achievers Know That Brilliant Machines Never Will*, New York: Penguin.

Davies, Alex (2015) "The World's First Self-Driving Semi-Truck Hits the Road," wired.com, May 5.

DeBord, Matthew (2015) "Tesla is in the Middle of a Huge Debate About the Future of Driving," businessinsider.com, September 13.

Denning, Steve (2015) "The 'Jobless Future' is a Myth," *Forbes*, June 4.

Dillon, Frank (2015) "The Personal Touch: Why Robots Will Not Put Us All Out of Our Jobs," *The Irish Times*, August 24.

Dwoskin, Elizabeth and Brian Fung (2016) "For Some Safety Experts, Uber's Self-Driving Taxi Test Isn't Something to Hail," *The Washington Post*, September 11.

Ford, Martin (2015) *Rise of the Robots: Technology and the Threat of a Jobless Future*, New York: Basic Books.

Frey, Carl Benedikt and Michael A. Osborne (2013) *The Future of Employment: How Susceptible are Jobs to Computerisation?*, Oxford: Oxford Martin School.

Frick, Walter (2014) "Experts Have No Idea If Robots Will Steal Your Job," *Harvard Business Review*, August 8.

Giang, Vivian (2015) "Robots Might Take Your Job, But Here's Why You Shouldn't Worry," fastcompany.com, July 28.

Haldane, Andy (2015) "Labour's Share," speech to the UK Trades Union Congress, London, September 15.

Hawkes, Rebecca (2016) "The Revenant: What Was Real and What Was Fake," *Telegraph*, January 20.

Hay, Mark (2015) "Why Robots are the Future of Elder Care," *Good*, June 24.

Hill, David J. (2015) "Kurzweil Responds to 'When Robots are Everywhere, What Will Humans be Good For?'," singularityhub.com, June 28.

Bibliography

Hoge, Patrick (2016) "Toyota Staffs Up for AI, Robotics Research," *Robotics Business Review*, March 1.

Insurance Times (2015) "Volvo to Accept Liability on Driverless Cars," *Insurance Times*, October 22.

Janoski, Thomas, David Luke, and Christopher Oliver (2014) *The Causes of Structural Unemployment: Four Factors that Keep People from the Jobs They Deserve*, Cambridge: Polity.

Kaplan, Jerry (2015) *Humans Need Not Apply: A Guide to Wealth and Work in the Age of Artificial Intelligence*, New Haven: Yale University Press.

Kennedy, Joe (2016) "Apple and America's Tax-Revenue Giveaway," entrepreneur.com, September 23.

Keynes, John Maynard (1931) "Economic Possibilities for Our Grandchildren," in *Essays in Persuasion*, London: Macmillan.

Kurzweil, Ray (2000) *The Age of Spiritual Machines: When Computers Exceed Human Intelligence*, London: Penguin.

Levy, Frank and Richard J. Murnane (2004) *The New Division of Labor: How Computers are Creating the Next Job Market*, Princeton: Princeton University Press.

Lloyd's (2014) "Autonomous Vehicles. Handing Over Control: Opportunities and Risks for Insurance," lloyds.com.

Marcus, Lucy (2015) "Is This a Truly Robot-proof Job?," bbc.com, September 21.

Moore, Andrew (2015) "Why 2016 Could be a Watershed Year for Emotional Intelligence in Machines," *Scientific American*, December 28.

Murray, Charles (2016) "A Guaranteed Income for Every American," *Wall Street Journal*, June 3.

Orfalea, Matt (2015) "Why Milton Friedman Supported a Guaranteed Income," medium.com, December 11.

Peterson, Hayley (2015) "The 12 Jobs Most at Risk of Being Replaced by Robots," weforum.org, November 2.

Bibliography

Ricardo, David (1821) *On the Principles of Political Economy and Taxation*, 3rd edition, London: John Murray.

Sainato, Michael (2015) "Stephen Hawking, Elon Musk, and Bill Gates Warn About Artificial Intelligence," *Observer*, August 19.

Samani, Kyle (2015) "Autonomous Cars Break Uber," techcrunch. com, September 18.

Santens, Scott (2015) "Self-Driving Trucks are Going to Hit Us Like a Human-Driven Truck," medium.com, May 14.

Saracco, Roberto (2016) "Guess What Requires 150 Million Lines of Code…", eitdigital.eu, January 13.

Schoettle, Brandon and Michael Sivak (2015) "Driverless Vehicles: Fewer Cars, More Miles," University of Michigan Transportation Research Institute.

Schumpeter (2015) "Professor Dr Robot QC," *The Economist*, October 17.

Schwartz, Samuel I. (2015) *Street Smart: The Rise of Cities and the Fall of Cars*, Philadelphia: PublicAffairs Books.

Smith, Aaron and Janna Anderson (2014) "AI, Robotics, and the Future of Jobs," pewinternet.org, August 6.

Stone, Maddie (2015) "The Trillion Fold Increase in Computing Power, Visualized," gizmodo.com, May 24.

Strauss, Steven (2015) "As the Age of Autonomous Vehicles Nears, Why are Policy Wonks Focused on the Past?", *Los Angeles Times*, November 6.

Summers, Lawrence H. (2013) "The 2013 Martin Feldstein Lecture: Economic Possibilities for Our Children," *NBER Reporter*, No. 4.

Susskind, Richard and Daniel Susskind (2015) *The Future of the Professions: How Technology Will Transform the Work of Human Experts*, Oxford: Oxford University Press.

Suster, Mark (2013) "In 15 Years from Now Half of US Universities May Be in Bankruptcy," bothsidesofthetable.com, March 3.

Bibliography

Technology Quarter (2016) "After Moore's Law," *The Economist*, March 12.

Thomis, Malcolm I. (1970) *The Luddites: Machine-Breaking in Regency England*, Newton Abbott: David and Charles.

Toffler, Alvin (1970) *Future Shock*, New York: Random House.

Vincent, James (2015) "Mercedes Thinks a Premium Driverless Taxi System Would Be a Good Idea," theverge.com, September 15.

Wharton (2015) "The Hype is Dead, But MOOCs are Marching On," knowledge.wharton.upenn.edu, January 5.

Wiener, Norbert (1950) *The Human Use of Human Beings*, Boston: Houghton Mifflin.

Zhenghao, Chen et al. (2015) "Who's Benefiting from MOOCS, and Why," *Harvard Business Review*, September 22.

ABS (antilock braking system) as example of robotic driver over-ride, 19
affective computing, 41
A.I. Artificial Intelligence (film, 2001), 104
AirBnB letting as new job, 76
Alexander, Bryan, futurist, 7
Apple, 97
  and self-driving cars, 20
Artificial General Intelligence, 73
ATMs, 79
Auerswald, Philip, critique of Martin Ford, 77–8
autonomous vehicles, *see* self-driving cars
Autor, David, "Why Are There Still So Many Jobs?", 48–9, 88, 90

Bear, The (film, 1988), 42
Bessen, James, *Learning by Doing*, 48
board membership and Machine Intelligence, 31–2
Boyd, Stowe, on leisure, 105
Brain, Marshall, x, 73
Brynjolfsson, Erik and Andrew McAfee, *The Second Machine Age*, 47–8

car ownership among millennials, 26
car ownership and self-driving vehicles, 25–8
car utilization in the US, 27–8
career choices and roboticization, 103
Carnegie Mellon University, 21
cascading effects of Machine Intelligence, 69–88
ChihiraAico, humanoid care companion, 38–9
Christensen, Clay, 33–5
coal-mining, decline of in the UK, 64
Colvin, Geoff, *Humans are Underrated*, 79
consensus, toward, 85–6, 87–109
Craigslist, launch of, 50

Daimler and self-driving cars, 28
DARPA (Defense Advanced Research Projects Agency), 57
deaths on the road, US, 23
demographics as factor in roboticization of elder care, 37
Detroit, Michigan, 20
  municipal bankruptcy, 65, 68
  and structural unemployment, 93

Index

digital revolution, economic impact of, 45–7

eBay selling as new job, 76
Economist, The, 45–6
education, 101–8
  root purpose of, 105–6
education and Machine Intelligence, *see* MOOCs
elder care and Machine Intelligence, 37–40
Elizabeth I and patent grant refusal, 17
European auto companies and self-driving cars, 21
European experiments with Universal Basic Income, 100–1

Facebook, 62–3
  launch of, 50
fallacy of the new normal, 53–8
fast food and roboticization, 96
financial services and Machine Intelligence, 29–32
Ford, Martin, *Rise of the Robots*, 48, 77
framing the question, importance of, 90–3
Frey, Carl Benedikt and Michael A. Osborne, 15–18, 30–1, 35–6, 40, 74
Friedman, Milton, and Universal Basic Income, 99
full-employment economy, risk of collapse, xiv, 3–4, 12, 59, 87, 95

Gary, Indiana, collapse of steel industry in, 65
  and structural unemployment, 93
Gates, Bill, 1, 6, 60, 75
Georgia Tech and MOOCs, 34
gig economy, 95
GMOs (Genetically Modified Organisms) and risk, 44, 81
Google, 50, 51, 54, 84, 97
  and self-driving cars, 18–20
GPS (Global Positioning System), 54

Haldane, Andy, chief economist, Bank of England, 65–6
Hargreaves' Spinning Jenny, 67
Harvard Business School, 33
Hawking, Stephen and AI warning letter, 45
high schools and colleges, avoidance of this discussion, 104
human quotient, as roboticization indicator, 37

IBM's Watson, 35–6
Industrial Revolution, xii, 11, 64, 66
Information Technology and Innovation Foundation, 110
insurance and self-driving vehicle safety, 22–5
Internet Explorer, launch of, 50

Japan and elder care, 37–9

Kaplan, Jerry, and car ownership, 26, 79

Index

Keynes, John Maynard, xiii, xiv, 2, 5, 72, 87
Kodak, 63
Kurzweil, Ray, 52, 73–5

labor leaders and risk, 92
labor market disruption, 4, 93–100
legal services and Machine Intelligence, 29–32
Lehman Brothers' collapse and risk, 57
leisure time, 102
and education, 105
Leonardo DiCaprio, 42
Levy, Frank and Richard Murnane on self-driving cars, 79
LinkedIn, 62
Lloyd's of London and self-driving cars, 24–7
Ludd, Ned, 7, 47, 110
Luddites, historical, xi
Luddites, metaphorical, 3, 5, 7, 9–11, 43–59, 92, 97

machine-breaking, 110
Marcus, Lucy, 31
Match.com, launch of, 50
Memorial Sloan Kettering Cancer Center, 36
minimum wage and roboticization, 96
mining companies and self-driving vehicles, 22
MIT, 9, 21, 43, 47, 84, 90
MOOCs (Massive Open Online Courses), 32–6, 108
Moore, Gordon, 51
Moore's Law, 50–7

Murray, Charles, xiv, 5, 71, 88
and UBI, 99–100
Musk, Elon as "Luddite of the Year", 44–5, 110
and AI warning letter, 45

natural-language recognition and use as key to care applications, 39
Netscape IPO, 50
new jobs, emergence of, 70–86
new leisured class, xiii
new machine species, 92
new normal, fallacy of, 53–8
new species, Machine Intelligence as, 63
nursing and Machine Intelligence, 37–40

Obama, Barack, 91
OECD (Organization for Economic Cooperation and Development), 17–18, 80, 88
and "full employment", 94
Oxford Martin Institute Report, see Frey, Carl Benedikt and Michael A. Osborne

Palro, robotic care companion, 38
parental leave, 95
Paro, robotic care companion, 38
Pepper, humanoid care companion, 39
Pew Research Center study, 5–7, 15
Picard, Rosalind and "affective computing", 41, 110
Pittsburgh, host to Uber cars, 1

Index

Polanyi, Michael, and tacit
    knowledge, 49
policy arena, absence of this
    discussion from, 104
policy issues, 93–100
policymakers, role of, 12–13, 87,
    90–3
    and the future of labor markets,
    59
    and industrial decline, 69
professionals and unemployment,
    106
professions and Machine
    Intelligence, 37, 50
psychiatry and Machine
    Intelligence, 40–1
public funding of science and
    technology, 57

recreational therapists, 15, 104
retirement, 95
Revenant, The (film, 2015), 42
Ricardo, David, xiv, 66–7
risk and planning, 89–109
risk and the future of employment,
    58–9
Roomba, early domestic
    application of Machine
    Intelligence, 38
Rust Belt, 3, 60–88

science fiction as educational tool,
    106
search engine optimization as new
    job, 76, 82
self-driving cars, 18–29
    and computer code, 19, 69
self-employment, 96

Singularity Institute, 73
Singularity University, 74
Skype, launch of, 50
slave economy as analog of robot
    economy, xiii, 14, 63
smoking and risk, 44
social devastation as consequence
    of structural unemployment,
    68
social security and Universal Basic
    Income, 100
social wage, potential European
    competitive advantage, 101
South Wales and structural
    unemployment, 93
Standage, Tom, 7
Stanford, 21, 26, 39, 84
STEAM (Science, Technology,
    Engineering, Arts, Math) as
    solution to roboticization,
    103
STEM (Science, Technology,
    Engineering, Math), 31,
    101–5
strategic planning approach to risk,
    89–109
structural employment, cost of,
    60–88, 87, 93
Summers, Larry, xiv, 5, 9–11, 43
survivability human quotient, 104
Susskind, Richard and Daniel
    Susskind, 37, 50, 60–1

tacit knowledge as factor in labor,
    49
Tamagotchi as precursor of more
    complex robot companions,
    38

Index

Tata and self-parking technology, 21

Teamsters, x

telemarketer, 15, 104

Titanic, sinking of, and risk, 44

Toffler, Alvin, *Future Shock*, 53

Toyota, 84
  investment in self-driving technology, 21

Trades Union Congress (TUC), UK, 65–6

training, 101–8

truck drivers and self-driving vehicles, 69

truck-driving jobs in the US, 22

Twitter, 62

Uber, 1, 18, 21, 28, 69, 70, 96–7

Uber driving as new job, 76

UBI, *see* Universal Basic Income

unemployment, 95
  social and political implications, 98–100
  long-term, 101

Universal Basic Income (UBI), 98–101
  as "socialist" or "conservative" solution, 99
  Milton Friedman as advocate, 99
  Charles Murray as advocate, 99–100
  European experiments with, 100

Universal Income, *see* Universal Basic Income

US Department of Labor, x, 15, 73

voluntary work, 102

Volvo and self-driving cars, 21
  and insurance, 24

Wall Street Collapse of 2008 and risk, 44

Welch, Jack, of General Electric, as leader, 56

Wiener, Norbert, xiii, 2, 5, 15, 83

workforce preparation, 101–8

World Bank, xi

World Economic Forum, 15, 31